# Incorporating Environmental Considerations into Defense Acquisition Practices

JEFFREY A. DREZNER, MEGAN MCKERNAN, GABRIEL LEONARD,
GWEN MAZZOTTA, SHANE MANUEL, SUSAN A. RESETAR,
SYDNE J. NEWBERRY

Prepared for the Office of the Secretary of Defense
Approved for public release; distribution unlimited

For more information on this publication, visit **www.rand.org/t/RRA2345-1**.

**About RAND**



**Research Integrity**



Published by the RAND Corporation, Santa Monica, Calif.


Library of Congress Cataloging-in-Publication Data is available for this publication.

ISBN: 978-1-9774-1137-2

*Cover: U.S. Marine Corps photo by Gunnery Sgt. Marcin Platek.*

**Limited Print and Electronic Distribution Rights**

# About This Report

The *2022 National Defense Strategy* (NDS) and recent National Defense Authorization Acts (NDAAs) emphasize the importance of adapting to the impacts of climate change as an element of national security. Accordingly, the 2022 NDS states on p. 2 that "We will make our supporting systems more resilient and agile in the face of threats that range from competitors to the effects of climate change."

In light of these challenges, the U.S. Congress required that the Office of the Under Secretary of Defense for Acquisition and Sustainment (OUSD[A&S]) respond to Section 873 of the fiscal year 2022 NDAA, which required an assessment of the knowledge, tools, and capabilities necessary for the acquisition workforce to infuse environmental considerations into U.S. Department of Defense (DoD) requirements, acquisition, and resource allocation decisionmaking. OUSD(A&S) asked the RAND Corporation's National Defense Research Institute to provide that independent analysis.

We used a policy and literature review and discussions with subject-matter experts to understand the current state of environmental practice in DoD acquisition and the knowledge and tools needed to incorporate environmental considerations in acquisition. In this report, we assume that the reader is at least somewhat familiar with DoD's environmental challenges, climate resiliency efforts, and defense acquisition policies and processes.

This report should be of interest to those concerned with DoD environmental initiatives, climate adaptation, and defense acquisition policy and practice. The intended audience includes members of Congress, congressional staff, and government officials responsible for environmental concerns, as well as the requirements for and acquisition of both weapon systems and goods and services.

The research reported here was completed in April 2023 and underwent security review with the sponsor and the Defense Office of Prepublication and Security Review before public release.

## RAND National Security Research Division

This research was sponsored by OUSD(A&S) and conducted within the Acquisition and Technology Policy Program of the RAND National Security Research Division (NSRD), which operates the National Defense Research Institute (NDRI), a federally funded research and development center sponsored by the Office of the Secretary of Defense, the Joint Staff, the Unified Combatant Commands, the Navy, the Marine Corps, the defense agencies, and the defense intelligence enterprise.

For more information on the RAND Acquisition and Technology Policy Program, see www.rand.org/nsrd/atp or contact the director (contact information is provided on the webpage).

## Acknowledgments

Within the Office of the Assistant Secretary of Defense (Acquisition)/Acquisition Enablers (ASD[A]/AE), we would like to thank the sponsor of this study, Mark Krzysko, Principal Deputy Director for Acquisition Data and Analytics (while also leading Enterprise Information), for his support. We would also like to thank David Cadman, the Acting Deputy Assistant Secretary of Defense, Acquisition Enablers, and Director of Acquisition Data and Analytics, for his support. In addition, Craig Scott Smith, Acting Director of Acquisition Approaches and Management, and Katherine Coyne, Middle Tier of Acquisition Pathway Lead within Acquisition Approaches and Management, provided valuable assistance in identifying and coordinating a long list of key acquisition stakeholders and their intellectual contributions. Likewise, we would like to thank David Asiello, Director of Sustainability & Acquisition within the Office of the Deputy Assistant Secretary of Defense (ODASD) (Environment & Energy Resilience), for his assistance in identifying and coordinating the key environmental stakeholders and for his thoughtful feedback throughout the analysis. Additionally, we acknowledge and thank the many DoD subject-matter experts who regularly met with us to help us further understand the data needed for this analysis: Carolyn Balven, Oliver Fritz, Sara Van Gorder, Everett (Skip) Hawthorne, Andrew Knox, Kevin Linden, Michael Mcghee, Larry McLaury, Kari Meier, Robin

Nissan, Michael Pelkey, Major Philip Song, Kimberly Spangler, Krista Stehn, and Patricia Underwood.

We also extend our thanks to officials from the following organizations that participated in this research by engaging in discussions to help us understand current methods for incorporating environmental considerations in acquisition:

- ASD(A)/AE
- ODASD (Environment & Energy Resilience)
- ODASD (Force Education & Training)
- OUSD(A&S)/Office of Defense Pricing and Contracting
- Office of the Assistant Secretary of the Air Force (Energy, Installations, and Environment)
- Office of the Assistant Secretary of the Army (Installations, Energy, and Environment)
- Office of the Assistant Secretary of the Navy (Energy, Installations, and Environment)
- U.S. Army Corps of Engineers
- Defense Acquisition University
- Air Force Institute of Technology
- Federal Acquisition Institute
- Naval Postgraduate School
- Headquarters Air Force Materiel Command/Contracting
- Air Force Life Cycle Management Center/Acquisition Environmental Integration Branch
- Army G-8/Program Analysis and Evaluation
- Office of the Assistant Secretary of the Army (Acquisition, Logistics, and Technology)
- Office of the Assistant Secretary of the Navy (Research, Development and Acquisition)
- U.S. Army Combat Capabilities Development Command.

Additionally, we thank our peer reviewers, William Shelton and Bruce Held, who helped improve the quality of this report through their comments and suggestions. We also thank Maria Falvo for her assistance during this effort. Finally, we would like to thank Christopher Mouton, acting director of RAND's Acquisition and Technology Policy Program, and Yun Kang, associate director of the Acquisition and Technology Policy Program, for their insightful comments on this research.

# Summary

## Issue

The impacts of climate change and other environmental threats are increasingly perceived by the White House and the U.S. Department of Defense (DoD) as a national security threat. Mitigating and adapting to those impacts is increasingly emphasized in national security policy.[1] In 2021, U.S. Secretary of Defense Lloyd Austin established a Climate Working Group to provide a DoD forum to coordinate responses and track implementation of climate- and energy-related initiatives throughout DoD.[2] DoD's acquisition community is a stakeholder in this forum because the impacts of climate change have wide-ranging implications for the acquisition and sustainment of weapon systems and combat support systems, and acquisition of goods and services. In light of these challenges, the U.S. Congress required that the Office of the Under Secretary of Defense for Acquisition and Sustainment (OUSD[A&S]) respond to Section 873 of the fiscal year 2022 National Defense Authorization Act (NDAA), which required an assessment of the knowledge, tools, and capabilities needed by the acquisition workforce to infuse environmental considerations into DoD requirements, acquisition, and resource allocation decisionmaking.[3] OUSD(A&S) asked the RAND

---

[1] Joseph R. Biden, "Executive Order on Tackling the Climate Crisis at Home and Abroad," Executive Order 14008, Executive Office of the President, January 27, 2021a; DoD, Climate 21 Project, homepage, 2021; DoD, *2022 National Defense Strategy of the United States of America*, October 27, 2022c; Jim Garamone, "DoD Office Focuses on Effects of Climate Change on Department," *Anchorage Press*, August 2, 2022; Mandy Mayfield, "SOFIC NEWS: Pentagon Looks to Incorporate 'Climate Resilience' into Future Weapon Systems," *National Defense*, May 19, 2021; Secretary of Defense, "Establishment of the Climate Working Group," memorandum for Senior Pentagon Leadership, Commanders of the Combatant Commands, and Defense Agency and DoD Field Activity Directors, U.S. Department of Defense, March 9, 2021; White House, *National Security Strategy*, October 12, 2022.

[2] Secretary of Defense, 2021.

[3] Public Law 117–81, National Defense Authorization Act for Fiscal Year 2022, Section 873, December 27, 2021.

Corporation's National Defense Research Institute to provide that independent analysis.

## Approach

The research included a policy and literature review, along with discussions with subject-matter experts to understand the current state of environmental practice in DoD acquisition, and the knowledge and tools needed to incorporate environmental considerations into acquisition. In this research, we examined the knowledge and tools available to the DoD acquisition workforce; we did not assess the sufficiency of DoD environmentally related activities relative to need or demand. Nor did we assess resource sufficiency (e.g., staffing levels, funding) of the various organizations within DoD that are responsible for addressing environmental compliance and performance.

## Key Findings

To incorporate environmental considerations across the variety of both environmental issues and acquisition activities identified in Section 873 of the fiscal year 2022 NDAA, which cover environmental management, impact, compliance, resilience, and adaptation, the DoD acquisition workforce needs policies and guidance that tell them with what and how to comply and where and how in the requirements and acquisition process to engage. The acquisition workforce needs resources (subject-matter experts [SMEs] and websites with information, useful links, models, and other tools) that provide information on the environmental and operational performance of systems, subsystems, components, and goods and services. The acquisition workforce also needs an understanding of technology that is currently in development that might mitigate environmental impacts (e.g., carbon emissions) or the impacts of the environment (e.g., extreme weather) on systems and goods and services. Finally, the acquisition workforce needs training and education on a variety of environmental considerations as part of their functional acquisition training.

In fact, these conditions currently exist in the acquisition community; DoD has been incorporating environmental considerations into acquisition planning and decisionmaking for at least several decades. Our key findings are as follows:

- DoD's acquisition workforce appears to have the knowledge and tools to incorporate environmental considerations into acquisition planning, practice, and decisionmaking for both weapon systems and goods and services. This knowledge is embodied in the SMEs that reside in the Office of the Secretary of Defense and armed services' energy, installation, and environment organizations. Tools include myriad resources accessible to the acquisition workforce via DoD, other federal agency, and partner organization websites.
- DoD has long-standing policy and guidance in many areas related to environmental compliance and impact; however, there is a potential gap in environmentally specific functional policy and guidance as applied to acquisition. These policies demonstrate awareness of the knowledge and tools needed to incorporate environmental considerations into acquisition planning and practice.
- For weapon systems, environmental considerations are incorporated in the systems engineering, design interface, and product support processes and treated like any other performance or compliance requirement.
- Policy on sustainability and Federal Acquisition Regulation Part 23 include rules and guidance for incorporating environmentally friendly preferences in the procurement of goods and services, including the source selection process. DoD leverages commercial and industry standards, which eliminates the need for DoD to separately verify the environmental performance of commercially available items.
- Participation in internal DoD forums (e.g., the Climate Working Group) and interagency forums enables the DoD acquisition community to be aware of and leverage knowledge and tools within DoD and other federal agencies. Participation in these forums also facilitates consistency in the application of environmental practices to acquisition.
- The DoD acquisition workforce has access to a variety of educational options and professional development to improve workforce knowledge of incorporating environmental considerations in acquisition.

Additional resources might be available in the future; the Climate Literacy Sub-Working Group is assessing climate literacy across the DoD workforce and will be recommending "means and methods for tailoring and/or improving climate education, training, and engagement."[4]

In summary, the knowledge and tools required to incorporate environmental cost, resource, and energy efficiency and resilience considerations exist in DoD's environmental organizations, and policy and regulations direct how and when those considerations should be input to acquisition planning and process generally and to source selection specifically. DoD also has several research and development initiatives that invest in and demonstrate environmentally friendly technologies and products.

## Recommendations

Our recommendations to improve the incorporation of environmental considerations into acquisition planning and decisionmaking **build on existing DoD environmental capacity, capabilities, and activities:**

- **Create a working-level environmental adviser function** to assist in the procurement of environmentally preferred goods and services. This would increase the visibility of DoD environmental policy and help achieve DoD's environmental goals. The environmental adviser, whether for weapon system acquisition or procurement of goods and services, should engage early in the requirements process where "design in" decisions can have a greater impact. For weapon systems, this function already exists at the system command level. For procurement of goods and services, the function could reside at the contracting office level.
- Along with maintaining and updating the environmental requirements in acquisition policy and the "how to" in existing functional

[4] Assistant Secretary of Defense for Readiness, *Climate Literacy Sub-Working Group Overview*, undated, p. 6. This briefing was provided to the authors by the Deputy Assistant Secretary of Defense (Force Education and Training) and is not available to the general public.

guidance, **create and maintain an environmental guidebook** as a resource for requirements developers, program managers, contracting officers, and others in the acquisition community. The intent is not to replace the guidance in existing functional policy and guidebooks but rather to bring together the knowledge and lessons of environmental management as applied in acquisition processes. Functional guidance is also easier to update than policy in response to changes in technology or environmental impacts that require changes in environmental practice. The Office of the Deputy Assistant Secretary of Defense (Environment and Energy Resilience) within the Assistant Secretary of Defense (Energy, Installations, and Environment) could work with the Office of Acquisition Enablers within the Assistant Secretary of Defense (Acquisition) and the services' energy, installation, and environment offices to develop the guidebook, which could then be posted on the Defense Acquisition University website alongside other acquisition-related functional guidebooks.

- **Continue and enhance collaboration and information-sharing across DoD** and with other federal, state, local, and industry organizations. This could include establishing a central repository of environmental performance data and other information, including lessons from past and ongoing initiatives and technology demonstrations.
- Finally, build on existing practices in the General Services Administration (GSA) and other federal contracting organizations by **establishing task order general contract vehicles with prequalified firms for select environmentally preferred goods and services** (in addition to what is already provided by GSA). These contract vehicles would most likely be indefinite delivery/indefinite quantity contracts organized around specific classes of commodities. Threshold levels of environmental performance can be set to reflect other federal agencies' environmental performance levels or could be unique to DoD when appropriate.

# Contents

# Figures and Tables

## Figures



## Tables

CHAPTER 1

# Introduction

## Background

The impacts of climate change are increasingly perceived by the White House and the U.S. Department of Defense (DoD) as a national security threat. Therefore, adopting policies to reduce DoD's contribution to climate change and the potential impacts of climate change on DoD activities is increasingly emphasized in national security policy.[1] For example, to prepare the DoD workforce to better enable climate-informed policy, DoD is adjusting its education and training programs through such initiatives as Deputy Secretary of Defense Kathleen Hicks' establishment of the Climate Literacy Sub-Working Group.[2] DoD is also examining the impacts of climate change on installations.[3] The Navy has experimented with alterna-

[1] Joseph R. Biden, "Executive Order on Tackling the Climate Crisis at Home and Abroad," Executive Order 14008, Executive Office of the President, January 27, 2021a; DoD, Climate 21 Project, homepage, 2021; Jim Garamone, "DoD Office Focuses on Effects of Climate Change on Department," *Anchorage Press*, August 2, 2022; DoD, *2022 National Defense Strategy of the United States of America*, October 27, 2022c; Mandy Mayfield, "SOFIC NEWS: Pentagon Looks to Incorporate 'Climate Resilience' into Future Weapon Systems," National Defense, May 19, 2021; Secretary of Defense, "Establishment of the Climate Working Group," memorandum for Senior Pentagon Leadership, Commanders of the Combatant Commands, and Defense Agency and DoD Field Activity Directors, U.S. Department of Defense, March 9, 2021.

[2] DoD, "DOD, Other Agencies Release Climate Adaptation Progress Reports," DoD News, October 6, 2022b.

[3] Office of the Under Secretary of Defense for Acquisition and Sustainment, *Report on Effects of a Changing Climate to the Department of Defense*, U.S. Department of

tive fuels for ships,[4] and DoD has explored alternative operational energy solutions, including recently experimenting with tactical electric vehicles.[5] Recognizing that the government should lead with respect to mitigating the potential impacts its policies could have on climate change and the potential impacts on DoD operations, facilities, infrastructure, and weapon systems, it is not unreasonable for DoD to require a long-term strategy for the acquisition and sustainment of weapon systems and combat support systems, the acquisition of commodities, and for the acquisition of services.

DoD has long understood that its activities and decisions affect the environment and its personnel (e.g., use of hazardous chemicals and water contamination, impacts of construction, land use management around installations). Environmental factors and weather conditions can affect force readiness through effects on weapons systems or equipment (e.g., design standards, maintenance requirements tied to anticipated operating conditions). The environment can and does affect its systems (e.g., moving aircraft from bases in the path of hurricanes).[6] Although environmental con-

---

Defense, January 10, 2019; U.S. Department of Defense Inspector General, *Evaluation of the Department of Defense's Efforts to Address the Climate Resilience of U.S. Military Installations in the Arctic and Sub-Arctic*, April 13, 2022.

[4] Gareth Evans, "US Green Fleet: A New Era of Naval Energy," *Naval Technology*, May 3, 2016; Christopher Frost, "The Great Green Fleet Operates in the South China Sea," *PACOM News*, March 4, 2016; "U.S. Navy Starts Alternative Fuel Use," *Maritime Executive*, January 20, 2016.

[5] Joe Saballa, "US Army Seeking All-Electric Vehicle Fleet to Slash Carbon Emissions," *Defense Post*, February 10, 2022.

[6] Other examples include the following:

- Changing water viscosity resulting from changes to ocean salinity and temperature affects underwater sensors used on submarines (Richard Nugee, "A Growing Crisis: The Launch of the World Climate and Security Report," Expert Group of the International Military Council on Climate and Security, June 7, 2021).
- "The stronger North Atlantic jet stream resulting from climate change will increase the risk of stronger wind shear and clear air turbulence. Since planes—in particular cargo planes—should avoid areas with strong turbulence, mission planning is further impaired" (Rene Heise, "NATO Is Responding to New Challenges Posed by Climate Change," *NATO Review*, April 1, 2021).

See also Katharina Ley Best, Scott R. Stephenson, Susan A. Resetar, Paul W. Mayberry, Emmi Yonekura, Rahim Ali, Joshua Klimas, Stephanie Stewart, Jessica Arana,

siderations are not a primary mission for DoD, DoD does have a history of complying with environmental regulations (e.g., the National Environmental Policy Act [NEPA]) and acknowledging that energy or environmental factors can affect its primary deterrent and combat missions (e.g., efficiency reduces the logistics fuel train; reducing the need for hazardous chemicals in system maintenance and operations improves occupational health and safety and, therefore, readiness). This research is intended to provide a strong foundation for understanding how DoD incorporates environmental considerations in acquisition decisionmaking and practice and, potentially, how to improve those processes.

The 2022 National Defense Strategy (NDS), the 2022 National Security Strategy, and recent National Defense Authorization Acts (NDAAs) emphasize the importance of adapting to the impacts of climate change as an element of national security. One objective of the 2022 NDS is to build a more "resilient Joint Force and defense ecosystem."[7] The intent is to adapt to "[c]hanges in global climate and other dangerous transboundary threats, including pandemics," that "are transforming the context in which the Department operates. We will adapt to these challenges, which increasingly place pressure on the Joint Force and the systems that support it."[8] As suggested in these high-level strategy and policy documents, DoD has two goals associated with incorporating environmental considerations into acquisition: (1) establishing practices and processes that minimize DoD's impacts on the environment and (2) making DoD and the national security appara-

---

Inez Khan, and Vanessa Wolf, *Climate and Readiness: Understanding Climate Vulnerability of U.S. Joint Force Readiness*, RAND Corporation, RR-A1551-1, 2023; Frank Camm, Jeffrey A. Drezner, Beth E. Lachman, and Susan A. Resetar, *Implementing Proactive Environmental Management: Lessons Learned from Best Commercial Practice*, RAND Corporation, MR-1371-OSD, 2001; Jeffrey A. Drezner and Melissa A. Bradley, *A Survey of DoD Facility Energy Management Capabilities*, RAND Corporation, MR-875-OSD, 1998; U.S. Department of Defense Environment, Safety and Occupational Health Network and Information Exchange (DENIX), *Defense Environmental Restoration Program Annual Report to Congress: Fiscal Year 1995*, 1995; and DENIX, *Calendar Year 2004: Executive Order 13148 Annual Report—Department of Defense*, May 16, 2005.

[7] DoD, *Fact Sheet: 2022 National Defense Strategy*, March 2022a, p. 1.

[8] DoD, 2022a, p. 1.

tus resilient to the impacts of climate change. This research addresses both goals.

Section 873 of the fiscal year (FY) 2022 NDAA (henceforth referred to as Section 873) mandates that DoD engage a federally funded research and development center (FFRDC) to study acquisition practices and policies to identify the knowledge and tools needed by the acquisition workforce to engage in planning practices that, in short, consider and promote environmental factors in acquisition decisions.[9] The Office of the Under Secretary of Defense for Acquisition and Sustainment (OUSD[A&S]) asked the RAND Corporation's National Defense Research Institute to provide that independent analysis.

The language used in Section 873 is broad in terms of both the kinds of acquisition-related activities that should be covered and the environmental considerations that need to be addressed:

> The study required under subsection (a) shall identify the knowledge and tools needed for the acquisition workforce of the Department of Defense to—
>
> (1) engage in acquisition planning practices that assess the cost, resource, and energy preservation difference resulting from selecting environmentally preferable goods or services when identifying requirements or drafting statements of work;
>
> (2) engage in acquisition planning practices that promote the acquisition of resilient and resource-efficient goods and services and that support innovation in environmental technologies, including—
>
> (A) technical specifications that establish performance levels for goods and services to diminish greenhouse gas emissions;
>
> (B) statements of work or specifications restricted to environmentally preferable goods or services where the quality, availability, and price is comparable to traditional goods or services;
>
> (C) engaging in public-private partnerships to design, build, and fund resilient, low-carbon infrastructure;

[9] Public Law 117–81, National Defense Authorization Act for Fiscal Year 2022, Section 873, December 27, 2021.

(D) collaborating with local jurisdictions surrounding military installations, with a focus on reducing environmental costs; and

(E) technical specifications that consider risk to supply chains from extreme weather and changes in environmental conditions;

(3) employ source selection practices that promote the acquisition of resilient and resource-efficient goods and services and that support innovation in environmental technologies, including—

(A) considering resilience, low-carbon, or low-toxicity criteria as competition factors on the basis of which the award is made in addition to cost, past performance, and quality factors;

(B) using accepted standards, emissions data, certifications, and labels to verify the environmental impact of a good or service and enhance procurement efficiency;

(C) evaluating the veracity of certifications and labels purporting to convey information about the environmental impact of a good or service; and

(D) considering the costs of a good or service that will be incurred throughout its lifetime, including operating costs, maintenance, end of life costs, and residual value, including costs resulting from the carbon dioxide and other greenhouse gas emissions associated with the good or service; and

(4) consider external effects, including economic, environmental, and social, arising over the entire life cycle of an acquisition when making acquisition planning and source selection decisions.[10]

Therefore, this research focused on training and education, policy and guidebooks, websites, subject-matter expertise, data, and analytic tools available to DoD's acquisition workforce. Two definitions included in Section 873 suggest the breadth and scope of activities to be included in the study:

(1) The term "environmentally preferable," with respect to a good or service, means that the good or service has a lesser or reduced effect on

[10] Public Law 117–81, Section 873, 2021.

> human health and the environment when compared with competing goods or services that serve the same purpose or achieve the same or substantially similar result. The comparison may consider raw materials acquisition, production, manufacturing, packaging, distribution, reuse, operation, maintenance, or disposal of the good or service.
>
> (2) The term "resource-efficient goods and services" means goods and services—
>
> > (A) that use fewer resources than competing goods and services to serve the same purposes or achieve the same or substantially similar result as such competing goods and services; and
> >
> > (B) for which the negative environmental impacts across the full life cycle of such goods and services are minimized.[11]

While much of the language in Section 873 refers to *goods and services*, which means commodity procurement and acquisition of services, the language also suggests weapon or combat system acquisition. We therefore defined the scope of acquisition activities addressed in this analysis to include

- acquisition of goods (commodities/supplies) and services
- adaptation of new and existing systems to environmental changes
- acquisition of weapon systems or technologies (research and development, procurement [manufacturing], operations and maintenance, and disposal).

The scope of these acquisition activities—planning, practices, and decisionmaking—also includes the implications of energy use and environmental impact across the acquisition life cycle, including research and development, procurement (manufacturing), operations and maintenance, and disposal.

In this research, we focused broadly on identifying the knowledge and tools needed by the acquisition community to incorporate environmental considerations into acquisition activities. We also attempted to document

[11] Public Law 117–81, Section 873, 2021.

some of DoD's past and current activities in the acquisition domain representing how the knowledge and tools related to environmental issues have been and are being applied. We explicitly did not evaluate whether those activities and initiatives are sufficient or well-designed and implemented.

In this report, we characterize the issues and challenges that DoD's acquisition workforce must address to incorporate environmental considerations into DoD requirements, acquisition, and resource allocation decisionmaking. We describe some of DoD's current initiatives toward that end and suggest ways to enhance DoD's ability to achieve that goal. This includes acquiring more environmentally friendly goods and services at DoD installations and incorporating more-efficient and more–environmentally friendly technologies in weapon and combat support systems.

## Study Objective and Approach

The objective of this study was to assist OUSD(A&S) with the response to Section 873 of the FY 2022 NDAA by conducting an independent assessment of the knowledge, tools, and capabilities necessary to include environmental considerations in DoD requirements, acquisition, and resource allocation decisionmaking.[12] This research focused on

- documenting and characterizing how DoD currently integrates environmental impact considerations into its acquisition activities
- identifying and assessing gaps in the knowledge and tools necessary to include environmental impact considerations in DoD acquisition.

Characterizing DoD's environmental policies, processes, and activities related to acquisition policy and process enables identification of the knowledge and tools required and whether there are any gaps in the knowledge and tools used. We expect to find such knowledge and tools in subject-matter experts (SMEs) residing in environmental organizations, acquisition and environmental policies and guidebooks, DoD and external websites,

[12] Public Law 117–81, Section 873, 2021.

and training and education initiatives and courses both within DoD educational institutions and accessible to the acquisition workforce through public and private education and training organizations.

We used a mixed-methods approach to address these tasks, which included a policy and literature review and semistructured discussions with various SMEs.

First, we reviewed relevant statutes, regulations, and policy (e.g., U.S. Environmental Protection Agency [EPA] regulations, executive orders [EOs], Federal Acquisition Regulation [FAR]/Defense Federal Acquisition Regulation Supplement [DFARS], and DoD policy). We then focused on official DoD documentation (e.g., Office of the Secretary of Defense [OSD] and service-level Climate Action Plans and environmentally focused office mission statements). To understand education and training opportunities, we consulted course catalogs, tools, and training from DoD and other federal educational institutions. We reviewed U.S. Government Accountability Office, FFRDC, and trade literature for analyses of environmental considerations in acquisition, along with examples of environmentally friendly technologies being introduced to DoD. Finally, we drew on an extensive catalog of acquisition and environmental policy research and analysis at RAND and other relevant documentation.[13]

Second, we developed a set of questions and tailored them to use in semistructured discussions to elicit information in approximately 20 discussions with more than 35 DoD, federal government, and FFRDC SMEs, as listed in Figure 1.1. The discussions can be separated according to the focus of the particular office involved: environment; education, training, and tools; contracting; and weapon systems policy and program life cycle. In many

[13] See, for example, Camm et al., 2001; Cynthia R. Cook, Éder Sousa, Yool Kim, Megan McKernan, Yuliya Shokh, Sydne J. Newberry, Kelly Elizabeth Eusebi, and Lindsay Rand, *Ensuring Mission Assurance While Conducting Rapid Space Acquisition*, RAND Corporation, RR-A998-1, 2022; Megan McKernan, Jeffrey A. Drezner, Mark V. Arena, Jonathan P. Wong, Yuliya Shokh, Austin Lewis, Nancy Young Moore, Judith D. Mele, and Sydne J. Newberry, *Using Metrics to Understand the Performance of the Adaptive Acquisition Framework*, RAND Corporation, RR-A1349-1, 2022; and Susan A. Resetar, Frank Camm, and Jeffrey A. Drezner, *Environmental Management in Design: Lessons from Volvo and Hewlett-Packard for the Department of Defense*, RAND Corporation, MR-1009-OSD, 1998.

**FIGURE 1.1**
**Completed Subject-Matter Expertise Discussions, by Focus Area**

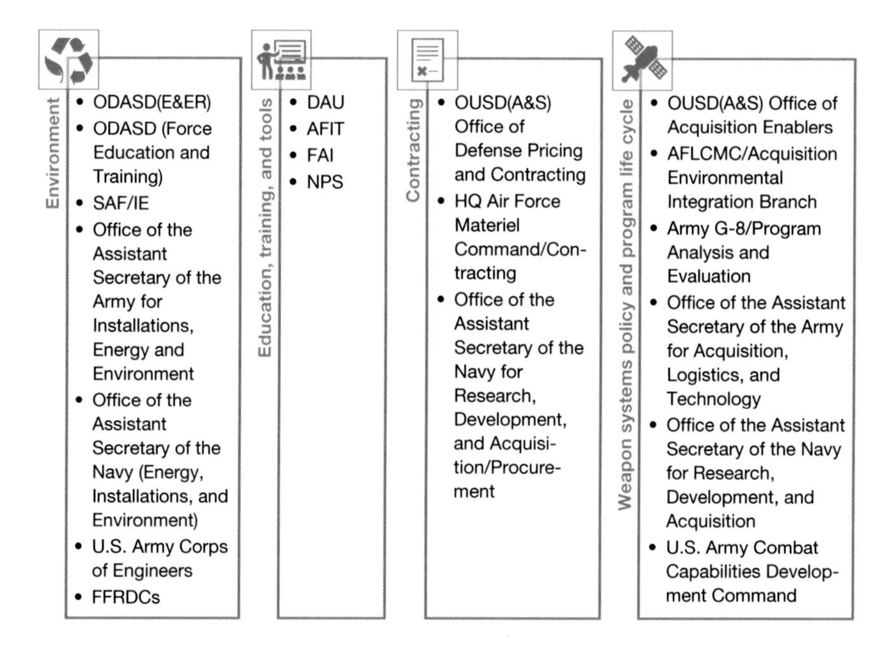


NOTE: Discussions covered only a portion of DoD activities because of the decentralization of activities throughout many organizations. AFIT = Air Force Institute of Technology. AFLCMC = Air Force Life Cycle Management Center. DAU = Defense Acquisition University. FAI = Federal Acquisition Institute. HQ = headquarters. NPS = Naval Postgraduate School. ODASD = Office of the Deputy Assistant Secretary of Defense. ODASD(E&ER) = Office of the Deputy Assistant Secretary of Defense (Environment and Energy Resilience). SAF/IE = Office of the Assistant Secretary of the Air Force for Energy, Installations, and Environment.

interviews with SMEs, additional recommendations for interviewees grew organically out of the discussion, pointing us to other organizations and individuals with expertise in the research topic. The discussions provided our analysis with a baseline understanding of the types of environmental impact consideration activities and training occurring in DoD's acquisition activities but were not meant to provide a comprehensive understanding of all environmental activities being undertaken within acquisition efforts across DoD. Because environmental and acquisition activities are decentralized across many organizations in DoD, discussions covered only a portion of DoD activities. As a result, and because of time and resource constraints, many relevant organizations and activities likely were not included in the interview process. Our intent was to cover enough activities to demonstrate whether DoD has the requisite knowledge and tools to incorporate environmental considerations in acquisition processes and how such knowledge and tools are used.

## Areas of Focus for Incorporating Environmental Considerations

DoD has been incorporating environmental considerations in defense acquisition for decades. This finding was corroborated through our many discussions with SMEs. Given this reality, it is important to acknowledge the types and scope of activities that have been occurring and to note that some level of overlap exists in these activities, particularly involving such factors as environmental considerations, preferences, and the safety concerns being addressed. This knowledge ultimately will provide a rough baseline for the purpose of understanding what knowledge and tools exist or might be missing in this space.

Although there are likely multiple ways to characterize the status quo, we chose to focus on several areas where DoD is incorporating environmental considerations. These areas are

- installations
- workforce health and safety
- weapon systems
- commercial goods and services.

In this chapter, we aim to provide some understanding of these four areas, but we do not assess the implementation progress in these areas.

### Installations

There are more than 500 military installations in the United States and overseas. Within these installations are more than 560,000 buildings and structures, according to the Office of the Deputy Assistant Secretary of Defense (Energy Directorate) (ODASD[Energy]).[14]

> The Department's inventory is diverse, encompassing barracks, commissaries, data centers, office buildings, laboratories, and aircraft maintenance depots. Installation energy consists largely of traditional energy sources used to heat, cool, and provide electrical power to these buildings. It also includes the fuel used by more than 160,000 non-tactical vehicles housed at DoD installations. The Department spends approximately $4 billion a year on energy that powers its fixed installations. Moreover, these bases are largely dependent on a commercial power grid that is vulnerable to disruption from aging infrastructure, weather-related events and direct attack.[15]

Given this large and diverse physical footprint, the focus of DoD environmental activities at installations and other facilities over time primarily has been on mitigating the impact of hazardous waste and toxic chemicals on personnel, reducing facility energy costs, and improving energy and water resilience while reducing environmental impacts within and around DoD bases and other related structures. More recently, DoD has been experimenting with the use of sustainable building materials to reduce overall environmental impact.

## Workforce Health and Safety

DoD employs approximately 3.4 million service members and civilians.[16] DoD focuses on protecting and safeguarding the health of its employees and their immediate families by adhering to laws, regulations, and policies at the federal, state, and local levels.

DoD also focuses on the health of the communities of people affected by the locations of DoD installations. Some methods of protecting people are through reducing the use of and exposure to toxic chemicals. DoD catego-

---

[14] Office of the Assistant Secretary of Defense for Sustainment, "Installation Energy," webpage, undated-b.

[15] Office of the Assistant Secretary of Defense for Sustainment, undated-b.

[16] Office of the Assistant Secretary of Defense for Sustainment, undated-b.

rizes some of the risks to human health through the Environmental, Safety, and Occupational Health (ESOH) framework,[17] which includes the following risks:

- "Hazardous Materials (HAZMAT) use and hazardous waste generation
- Safety (including explosives safety, radiation, etc.)
- Human health (chemical, physical, biological, ergonomic, etc.)
- Environmental and occupational noise
- Impacts to the environment (air, water, soil, flora, fauna)."[18]

## Weapon Systems

Development of DoD weapon systems is primarily focused on achieving operational capabilities, but a variety of environmental considerations are part of programmatic and technical decisions, including energy supportability and demand reduction. Energy efficiency is a Key Performance Parameter (KPP) as required by law for new systems and upgrades:

> (b) Capability Requirements Development Process.—The Secretary of Defense shall develop and implement a methodology to enable the implementation of a fuel efficiency key performance parameter in the requirements development process for the modification of existing or development of new fuel consuming systems.[19]

---

[17] DoD policy for the defense acquisition system is to emphasize ESOH risks and requirements management. ESOH is

> the combination of disciplines that encompass the processes and approaches for addressing laws, regulations, Executive Orders (EO), DoD policies, environmental compliance, and hazards associated with environmental impacts, system safety (e.g., platforms, systems, system-of-systems, weapons, explosives, software, ordnance, combat systems), occupational safety and health, hazardous materials management, and pollution prevention. (Military Standard 882E, *Department of Defense Standard Practice: System Safety*, U.S. Department of Defense, May 11, 2012)

[18] DAU, "Systems Engineering Brainbook: ESOH Risk Assessment," webpage, undated-e.

[19] U.S. Code, Title 10, Section 2911, Energy Policy of the Department of Defense: Statutory Notes and Related Subsidiaries—Consideration of Fuel Logistics Support Require-

Also, Department of Defense Instruction (DoDI) 5000.02 states that "[i]n consultation with the user representative, the [program manager] will determine which environment, safety, and occupational health risks must be eliminated or mitigated, and which risks can be accepted."[20] Program managers focus on an evaluation of a program's safety through the Programmatic Environmental, Safety, and Health Evaluation (PESHE), which is

> an overall evaluation of a programs [sic] Environmental, Safety and Occupational Health (ESOH) risks and is required for all programs. The PESHE is part of a program's Risk Management/Reduction program with the goal to eliminate ESOH hazards, where possible, and manage their associated risks where hazards cannot be eliminated.[21]

## Commercial Goods and Services

DoD buys commercial goods and services for many purposes (e.g., copy paper, paint, building materials, computers). In these procurements, DoD is guided by FAR 23.103, Sustainable Acquisitions, which was enacted in 2009 and states that "[f]ederal agencies shall advance sustainable acquisition by ensuring 95% of new contract actions for the supply of products and for the acquisition of services (including construction), are environmentally preferable products or services."[22]

---

ments in Planning, Requirements Development, and Acquisition Processes, Pub. L. 110–417, [div. A], title III, §332, Oct. 14, 2008, 122 Stat. 4420, as amended by Pub. L. 111–383, div. A, title X, §1075(e)(5), Jan. 7, 2011, 124 Stat. 4374.

[20] Department of Defense Instruction 5000.02, *Operation of the Adaptive Acquisition Framework*, U.S. Department of Defense, January 23, 2020, p. 11.

[21] AcqNotes, "Program Management: Programmatic Environmental Safety and Occupational Health Evaluation," webpage, June 14, 2018.

[22] Federal Acquisition Regulation, Part 23, Environment, Energy and Water Efficiency, Renewable Energy Technologies, Occupational Safety, and Drug-Free Workplace, Subpart 23.103, Sustainable Acquisitions, March 16, 2023b.

## Organization of This Report

In Chapter 2, we identify the DoD organizations in which environmental subject-matter expertise resides and that play an important role in setting environmental policy and incorporating environmental considerations in acquisition. These organizations have, maintain, and use knowledge and tools to inform acquisition decisionmaking. Chapter 3 focuses on environmental activities related to acquisition that demonstrate some of the knowledge and tools available to the acquisition workforce, including routine DoD environmentally related reports, research and development activities, and discussions of how DoD leadership acknowledges excellence in environmental practices. These activities demonstrate how DoD addresses environmental considerations in acquisition planning with respect to cost, resources, and energy efficiency and resilience, along with investment in environmentally friendly technologies. In Chapter 4, we examine acquisition and environmental policy and processes for various types of acquisitions. The information in this chapter demonstrates how DoD includes environmental considerations in source selection and how it addresses environmental compliance and impact. Chapter 5 focuses on educational and training opportunities, along with other resources available to the acquisition workforce. Finally, in Chapter 6, we present our key findings, recommendations, and areas for further research.

CHAPTER 2

# U.S. Department of Defense Environmental Organizations

## DoD's Organizational Structure for Incorporating Environmental Considerations Is Relatively Decentralized

Several organizations and directorates across OSD and the services are charged with managing departmental environmental considerations. It is these organizations that have the environmental knowledge (SMEs) and an understanding of associated tools, liaise with other agencies to maintain that expertise, and act as environmental advisers to decisionmakers. But, because DoD has a relatively decentralized and stovepiped organizational structure, interface with the acquisition community is not straightforward. DoD-wide policy authority on environmental issues resides in the Office of the Deputy Assistant Secretary of Defense (Environment and Energy Resilience) (ODASD[E&ER]) within the Assistant Secretary of Defense (Energy, Installations, and Environment).[1] Each of the services that the team consulted for this study—the Army, Air Force, and Navy—operates separate offices for managing environmental affairs (which we describe in further detail below) that are also outside the services' acquisition organizations. Although there are slight differences among the services in terms of structure, reporting mechanisms, and publicly available information on their

[1] This office recently moved from the Office of the Assistant Secretary of Defense for Sustainment to the Assistant Secretary of Defense (Energy, Installations, and Environment) (OUSD[A&S], "OUSD A&S Organizations," webpage, undated-b).

websites, the scope of their environmental responsibilities is generally similar and very broad. The services' environmental responsibilities typically deal with issues including safety, occupational health, hazardous materials, waste management, and restoration. Using material gathered from these offices' websites and through discussions with SMEs, we found evidence of significant collaboration across organizations, despite each service branch being granted a considerable amount of discretion in implementing environmental regulations and directives.

## Office of the Secretary of Defense

Authority on environmental and acquisition issues in OSD resides in OUSD(A&S). Within OUSD(A&S), the environmental offices reside in the Office of the Assistant Secretary of Defense (Energy, Installations, and Environment) (ASD[EI&E]), while the acquisition offices reside in ASD (Acquisition). ASD(EI&E) serves broadly as a principal adviser to OUSD(A&S) on energy, installation, and environment matters.[2] As of April 2023, the ODASD(E&ER) is responsible for overseeing environmental regulatory compliance and environmental restoration efforts and for providing policy direction on both the department's operational energy and installation energy resilience efforts and environmental issues more broadly. Other responsibilities include providing strategic direction to climate change adaptation and organizational sustainability initiatives and managing more than $250 million in research funds to advance environmental and energy technologies.[3] ODASD(E&ER) oversees the Strategic Environmental Research and Development Program (SERDP) and the Environmental Security Technology Certification Program (ESTCP).[4] It also oversees the Armed Forces Pest Management and Defense Explosives Safety boards and maintains an Energy, Installations, and Environment Library that contains

[2] OUSD(A&S), "Brendan Owens: Assistant Secretary of Defense for Energy, Installations, and Environment," webpage, undated-a.

[3] Office of the Assistant Secretary of Defense for Sustainment, "Deputy Assistant Secretary of Defense for Environment & Energy Resilience," webpage, undated-a.

[4] SERDP and ESTCP, "About Us," webpage, undated-b.

an archive of environmentally related testimonies, reports to Congress, and DoD internal reports.[5]

Within the Office of the Under Secretary of Defense for Personnel and Readiness, the Deputy Assistant Secretary of Defense for Force Education and Training "serves as the principal senior authority on the development of DoD policy on all issues related to military education and training across the Joint Force,"[6] which includes a "Climate Literacy portfolio [that] focuses on empowering military and civilian personnel with the education, training, and information that they require to execute their missions, protect our nation, and return home safely, no matter the environment."[7] Of note is the Climate Working Group and, specifically, the Climate Literacy Sub-Working Group (CLSWG) authorized by Deputy Secretary of Defense (DEPSECDEF) Hicks in January 2022. DEPSECDEF Hicks "tasked the CLSWG to develop a plan to incorporate climate considerations into DoD education and training programs to support a climate-literate workforce."[8] The CLSWG mission is to build, develop, and maintain a climate-literate workforce that actively addresses the impacts of climate change on DoD operations, helps DoD meet and overcome evolving climate risks, and makes lasting reductions to DoD's carbon pollution footprint. The CLSWG is chaired by the Deputy Assistant Secretary of Defense for Force Education and Training.

---

[5] Office of the Assistant Secretary of Defense for Sustainment, "Library, Resources & Archives," webpage, undated-c; Office of the Assistant Secretary of Defense for Sustainment, "Welcome to Energy," webpage, undated-d; Office of the Assistant Secretary of Defense for Sustainment, "Welcome to Environment," webpage, undated-e.

[6] Deputy Assistant Secretary of Defense for Force Education and Training, "About Us," webpage, undated-a.

[7] Deputy Assistant Secretary of Defense for Force Education and Training, "Programs," webpage, undated-b.

[8] Assistant Secretary of Defense for Readiness, "Climate Literacy Sub-Working Group Overview," undated, p. 4. This briefing was provided to the authors by the Deputy Assistant Secretary of Defense (Force Education and Training) and is not available to the general public.

## Department of the Army

The Office of the Assistant Secretary of the Army (Installations, Energy, and Environment) (ASA[IE&E]) is the focal point for the Army's environmental policy. Reporting directly to the Secretary of the Army, ASA(IE&E) is responsible for "establish[ing] policy, provid[ing] strategic direction and supervis[ing] all matters pertaining to infrastructure, Army installations and contingency bases, energy, and environmental programs to enable global Army operations."[9]

ASA(IE&E) manages several programs and produces plans, strategies, and directives to help Army systems incorporate environmental considerations. Key documents include the 2020 *Army Climate Resilience Handbook*,[10] the 2020 *Army Installation Energy and Strategic Water Plan*,[11] Army Directive 2020-11: *Roles and Responsibilities for Military Installation Operations*,[12] Army Directive 2020-03: *Installation Energy and Water Resilience Policy*,[13] and the *Safety, Occupational and Environmental Health (SO&EH) Strategy 2020–2028*.[14] For FY 2021, major accomplishments include signing the Environmental Justice policy, which was coordinated by the Office of the Deputy Assistant Secretary of the Army for Energy and Sustainability (ODASA [Energy & Sustainability]) team, and convening the

---

9 ASA(IE&E), "About Us" webpage, undated-a.

10 A. O. Pinson, K. D. White, S. A. Moore, S. D. Samuelson, B. A. Thames, P. S. O'Brien, C. A. Hiemstra, P. M. Loechl and E. E. Ritchie, *Army Climate Resilience Handbook*, U.S. Army Corps of Engineers, August 2020.

11 U.S. Army, *Army Installation Energy and Water Strategic Plan*, December 2020.

12 Army Directive 2020-11, *Roles and Responsibilities for Military Installation Operations*, U.S. Department of Defense, October 1, 2020.

13 Army Directive 2020-03, *Installation Energy and Water Resilience Policy*, U.S. Department of Defense, March 31, 2020.

14 Office of the Assistant Secretary of the Army for Installations, Energy and Environment, *Safety, Occupational and Environmental Health (SO&EH) Strategy 2020–2028*, U.S. Army, April 28, 2020. See Assistant Secretary of the Army (Installations, Energy and Environment), "Helpful Links to ASA(IE&E) Directorates, Programs and the Installation Management Community," webpage, undated-b.

Army Climate Change Working Group, which drafted the Army Climate Strategy and Army Climate Action Plan.[15]

Three Deputy Assistant Secretary offices—whose responsibilities include providing policies, programming, and oversight for Army programs related to their respective titles—report to ASA(IE&E):

- ODASA (Environment, Safety & Occupational Health)
- ODASA (Installations, Housing & Partnerships)
- ODASA (Energy & Sustainability).

In addition to providing policy, programming, and oversight, ODASA (Environment, Safety & Occupational Health) provides input into the acquisition process by advising milestone decision authorities on ESOH issues.[16]

The Principal Deputy Assistant Secretary of the Army (Installations, Energy and Environment) manages two offices: Strategic Integration, which is related to installation planning and development,[17] and the Army Climate Directorate, which is the principal organization that deals with actions and activities outlined in the Army Climate Strategy and Implementation Plan.

The office also publishes an annual ASA(IE&E) Year in Review that documents significant accomplishments from the prior year.[18]

## Department of the Air Force

SAF/IE is the focal point for the Department of the Air Force's environmental policy. SAF/IE is a direct report to the Under Secretary of the Air Force.[19]

[15] Office of the Assistant Secretary of the Army for Installations, Energy and Environment, *ASA (IE&E): Installations, Energy and Environment—Fiscal Year 2021: Year in Review*, U.S. Army, October 1, 2021.

[16] Office of the Assistant Secretary of the Army for Installations, Energy and Environment, "ESOH," webpage, undated-b.

[17] Office of the Assistant Secretary of the Army for Installations, Energy and Environment, "Strategic Integration," webpage, undated-c.

[18] Office of the Assistant Secretary of the Army for Installations, Energy and Environment, 2021.

[19] U.S. Department of the Air Force organizational chart, AFVA 38-104, supersedes AFVA 38-104, November 16, 2022.

The team comprises military, civilian, and contractor personnel who oversee and implement policies related to ESOH, operational energy, and installations (and their energy use). More broadly, part of SAF/IE's stated responsibilities includes managing all aspects of programs that sustain the Air Force's mission in light of potential disruptions or constraints caused by federal or state legislation and regulations related to energy, environment, infrastructure, installations, or safety.[20] The office also engages in cross-agency and cross-government partnerships with organizations including OSD, Congress, the administration, and major commands.[21]

SAF/IE manages three directorates:

- SAF Installations (SAF/IEI)
- SAF Environment, Safety, and Infrastructure (SAF/IEE)
- SAF Energy Program.

Each of the three offices offers tools and resources to help inform Air Force personnel at multiple levels.[22] SAF/IEI is responsible for supporting personnel and logistical combat capability, including military construction, asset management, base closures, family housing, community and congressional interface, joint use of airfields, and disposal of real property.[23]

SAF/IEE manages such activities as water resource management, policy development, environmental compliance, natural and cultural resource management, and safety, including radiation safety and radioactive materials management.[24] It also manages the Installation Energy program, which is one of three branches of the SAF/IE Energy Program, which we discuss below.

Finally, the SAF/IE Energy Program supports the Air Force's vision of "Mission Assurance through Energy Assurance" while improving cost-

---

[20] SAF/IE, "About Us," webpage, undated-a.

[21] SAF/IE, undated-a.

[22] SAF/IE, "Installations," webpage, undated-d.

[23] SAF/IE, undated-d.

[24] SAF/IE, "SAF/IEE Environment, Safety, and Infrastructure: What We Do," webpage, undated-f.

effectiveness and energy resiliency.[25] The Energy Program contains three supporting subprogram areas: Installation Energy, the Office of Energy Assurance, and Operational Energy (SAF/IEN).[26] The Installation Energy office ensures resilient energy and water systems in Air Force infrastructure. The Office of Energy Assurance operates multidisciplinary teams that identify opportunities for energy and water infrastructure improvements that advance the mission and align with Department of the Air Force Installation Energy Strategic Plan's goals.[27] SAF/IEN aims to increase efficiency and decrease fuel supply chain vulnerabilities while ensuring the Air Force's warfighting mission: "Our goal is to fly *smarter*, not less."[28]

SAF/IE plays a key role in informing Air Force–wide environmental initiatives. In October 2022, it published the Department of the Air Force Climate Action Plan, which includes such priorities as incorporating climate considerations into department professional military, technical, and continuing education curricula by FY 2024.[29]

There is some evidence of publicly available annual reviews or similar processes being conducted for SAF/IE departments, but efforts to either conduct or publish such reviews appear inconsistent; SAF/IEN most recently published an Annual Report in 2021 (for the year 2020),[30] and SAF/IEE's most recent publicly accessible Year in Review was published in December 2018 (for that year).[31]

---

[25] SAF/IE, "Air Force Energy Program," webpage, undated-b.

[26] SAF/IE, undated-b.

[27] SAF/IE, "Office of Energy Assurance: Your Storefront for Creative Energy Solutions," website, undated-e.

[28] SAF/IE, "Air Force Operational Energy About Us," webpage, undated-c.

[29] Department of the Air Force, Office of the Assistant Secretary for Energy, Installations, and Environment, *Climate Action Plan*, October 2022.

[30] Air Force Operational Energy, *Annual Report 2020*, Office of the Assistant Secretary of the Air Force for Energy, Installations, and Environment, undated.

[31] Melissa Tiedeman, "Year in Review: 2018 SAF/IEE Installation Energy," Office of the Assistant Secretary of the Air Force for Energy, Installations, and Environment, December 20, 2018.

## Department of the Navy

The Navy's focal point for environmental policy is the Assistant Secretary of the Navy (Energy, Installations, and Environment) (ASN[EI&E]), which is a direct report to the Under Secretary of the Navy. ASN(EI&E)'s mission is to enhance combat capabilities for the warfighter through

- managing installations/ranges and mitigating compatibility risks to protect and preserve Department of the Navy equities
- aligning climate actions to strengthen maritime dominance, empower people, and strengthen strategic partnerships
- increasing energy security, construction, and maintenance of installations
- protecting the safety and occupational health of military and civilian personnel
- protecting the environment
- planning and restoration ashore and afloat, conserving natural and cultural resources, and integrating environmental considerations in acquisition strategies.

ASN(EI&E) manages three directorates:

- the Deputy Assistant Secretary of the Navy (Installations, Energy and Facilities) (DASN[IE&F])[32]
- the Deputy Assistant Secretary of the Navy (Environment and Mission Readiness) (DASN[E&MR])[33]
- the Deputy Assistant Secretary of the Navy (Safety) (DASN[Safety]).[34]

DASN(IE&F) oversees issues including real estate, encroachment, housing, and infrastructure, and it provides a variety of policy and planning resources. DASN(E&MR) serves as the principal adviser to ASN(EI&E) for all matters pertaining to environmental protection, compliance, restora-

---

[32] ASN(EI&E), "Deputy Assistant Secretary of the Navy (Installations, Energy and Facilities)," webpage, undated-b.

[33] ASN(EI&E), undated-b.

[34] ASN(EI&E), "Deputy Assistant Secretary of the Navy (Safety)," webpage, undated-c.

tion, and technology, and management of natural, historical, and cultural resources for Department of Navy activities worldwide. The office coordinates with organizations outside DoD, such as the President's Council on Environmental Quality, the Environmental Protection Agency, and nongovernmental organizations (NGOs).[35] Its program elements, which cover issues including environmental restoration, quality, technology, and historical preservation, have a budget of more than $1 billion annually.[36] DASN(Safety) focuses on matters related to risk management, occupational safety and health, industrial hygiene, and similar areas.

ASN(EI&E) most recently published an annual report detailing FY 2020 accomplishments, priorities, and financials, but it has not released any other readily available annual reports.[37]

---

[35] ASN(EI&E), "Deputy Assistant Secretary of the Navy (Environment and Mission Readiness)," webpage, undated-a.

[36] ASN(EI&E), undated-a.

[37] ASN(EI&E), *Fiscal Year 2020 Annual Report*, Department of the Navy, undated.

CHAPTER 3

# Environmental Activities

DoD engages in several environmental activities that are part of or relate to acquisition processes. Additionally, each of these activities implicitly demonstrates the use of knowledge and tools needed to support an activity.

## DoD Routinely Reports on Environmental Activities

DoD routinely provides several annual reporting documents to Congress and the public on environmental issues.[1] These reports summarize information on department activities, initiatives, and metrics covering energy and environmental issues. Examples of this reporting include annual DoD sustainability plans; energy and sustainability performance reporting; defense environmental programs' annual reporting to Congress; DoD's climate adaptation plans; and service-level sustainability and climate plans, year in review documents, and initiatives.

DoD sustainability plans are annual reports that present (1) departmental priorities and strategies to meet goals; (2) initiatives; (3) progress improvements; and (4) mission improvements that relate to logistics and resupply, installation and range management, as well as operations, mission requirements, and resiliency. DoD's first plan was issued in 2010 in response to EO 13514, *Federal Leadership in Environmental, Energy, and Economic*

---

[1] Recent EOs have also added reporting requirements (e.g., Joseph R. Biden, "Executive Order on Catalyzing Clean Energy Industries and Jobs Through Federal Sustainability," Executive Order 14057, Executive Office of the President, December 8, 2021b). See also Office of the Federal Chief Sustainability Officer, Council on Environmental Quality, "Federal Progress, Plans, and Performance," webpage, undated-b.

*Performance*.[2] The content of these plans focuses on improving military readiness through resilient infrastructure, cost reduction, and personnel health and safety. The FY 2022 plan presents activities and progress toward sustainability goals that include 100-percent carbon pollution–free electricity by 2030; 100-percent zero-emissions vehicle acquisitions; net-zero emissions and increased energy and water efficiency of buildings, campuses, and installations; reductions in waste and pollution; and sustainable procurement. Examples of each of these goals are presented in DoD's FY 2022 plan. Notably, the FY 2016 plan stated that DoD reviewed FY 2015 contract actions and determined that more than 96 percent complied with the sustainable procurement requirement before DoD's sustainable procurement policy (DoDI 4105.72) was issued.[3]

The U.S. Office of Management and Budget (OMB) is required by law to publish an annual scorecard on agency energy efficiency and sustainability performance; the scorecard includes metrics for a variety of performance criteria, such as facility energy efficiency, renewable energy use, efficiency measures and investment, water use, high-performance sustainable buildings, greenhouse gas emissions, and fleet management. The sustainable acquisition metrics reported in the scorecard are the number and value of applicable contract actions with sustainable requirements clauses.[4] DoD also provides its energy and sustainability performance reporting as part of this requirement and provides the time series of its annual energy and sus-

---

[2] Office of the President, "Executive Order 13514—Federal Leadership in Environmental, Energy, and Economic Performance," *Federal Register*, Vol. 74, No. 194, October 8, 2009.

[3] DoDI 4105.72, *Procurement of Sustainable Goods and Services*, U.S. Department of Defense, September 7, 2016, change 1, August 31, 2018. See also DoD, *Department of Defense Strategic Sustainability Performance Plan*, Under Secretary of Defense for Acquisition, Technology, and Logistics, August 26, 2010; DoD, *Department of Defense Strategic Sustainability Performance Plan: FY 2016*, Under Secretary of Defense for Acquisition, Technology, and Logistics, September 7, 2016; and OUSD(A&S), *Department of Defense Sustainability Plan: 2022*, U.S. Department of Defense, March 2022b.

[4] OMB, *Department of Defense FY2021 OMB Scorecard for Federal Sustainability*, 2022.

tainability performance data for the period 2010 to 2021 on its sustainability website.[5] See Figure 3.1 for the reports provided annually.

The *Defense Environmental Programs Annual Report to Congress* provides information on the program's annual funding and status reports for environmental quality, restoration, and technology development.[6] The most recent report, which covered activities and funding in FY 2020, was provided in March 2022. Specifically,

> [t]he Report describes the Department's accomplishments during the past year in its restoration, conservation, compliance, and pollution prevention programs by addressing plans and funding needs for protecting human health, sustaining the resources DoD holds in the public trust, meeting its environmental requirements, and supporting the military mission. The Report also details DoD's efforts for reinforcing environmental programs to ensure the safe and effective use, protection, restoration, and preservation of the Department's natural and cultural assets; and examines DoD's environmental restoration activities at sites on its active and Base Realignment and Closure (BRAC) installations and former properties.[7]

EO 14008, *Tackling the Climate Crisis at Home and Abroad* and EO 14057, *Catalyzing Clean Energy Industries and Jobs Through Federal Sustainability* both require that DoD create climate adaptation plans. These annual plans (shown in Figure 3.2) are a means to communicate what DoD is doing to bolster climate adaptation and resilience.[8] For each of the lines of effort (LOEs) in the *2021 Climate Adaptation Plan*, DoD provides a description of the continuing effort; focus areas; agency leads; time frame; intergovernmental coordination; and potential risks, opportunities, performance metrics, and resource implications. Many of these activities relate to acquisition and sustainable procurement.

---

[5] Office of the Federal Chief Sustainability Officer, Council on Environmental Quality, "Department of Defense Agency Progress," webpage, undated-a.

[6] DENIX, "Annual Reports to Congress," webpage, undated-b.

[7] DENIX, undated-b.

[8] DoD, 2022b.

**FIGURE 3.1**

**Defense Sustainability Planning and Environmental Programs Annual Reporting Documents and OMB Scorecard**

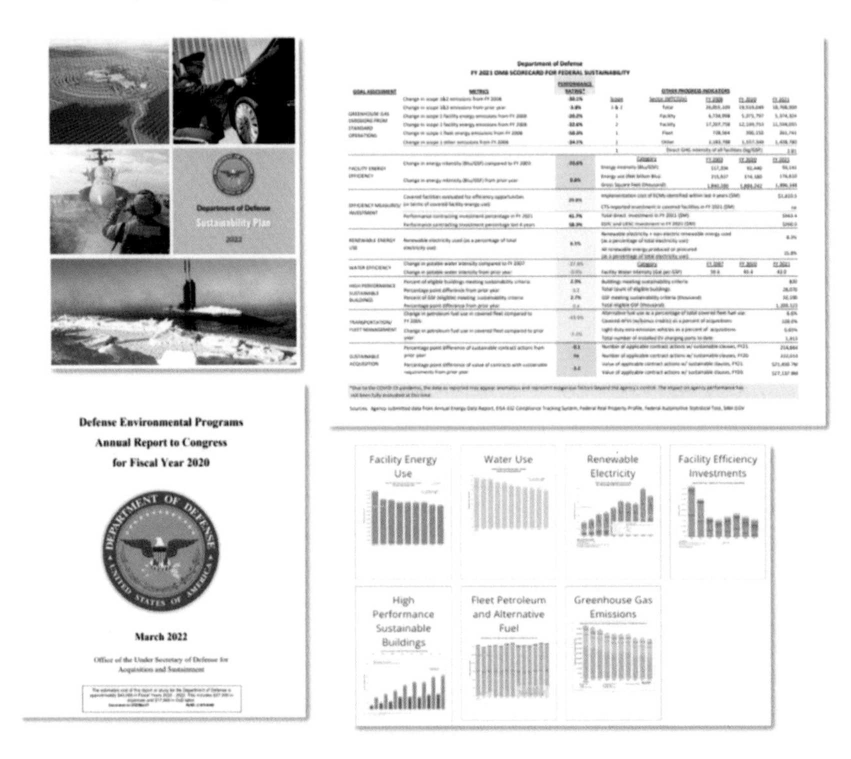


SOURCE: Reproduced from OUSD(A&S), 2022b; OMB, 2022; OUSD(A&S), *Defense Environmental Programs Annual Report to Congress for Fiscal Year 2020*, March 2022a; and Office of the Federal Chief Sustainability Officer, Council on Environmental Quality, undated-a (in order from top left to bottom right).

DoD's climate adaptation plan progress report provides a brief overview of current activities and examples for each LOE:

- LOE 1: Climate-Informed Decision-Making
- LOE 2: Train and Equip a Climate-Ready Force
- LOE 3: Resilient Built and Natural Installation Infrastructure
- LOE 4: Supply Chain Resilience and Innovation

- LOE 5: Enhance Adaptation and Resilience Through Collaboration.[9]

The report also includes information on special topics, such as climate scenario analysis and risk-reduction activities, vulnerability assessments, workforce literacy, tribal engagement, environmental justice, policy reviews, and partnerships. Finally, the report presents the organizational structure used to address climate issues.[10]

The services also report on their own sustainability and climate plans and associated initiatives, although there is no DoD requirement to do so.[11]

**FIGURE 3.2**
**Newer Initiatives Outlined in DoD Climate Adaptation Plans or Strategies**

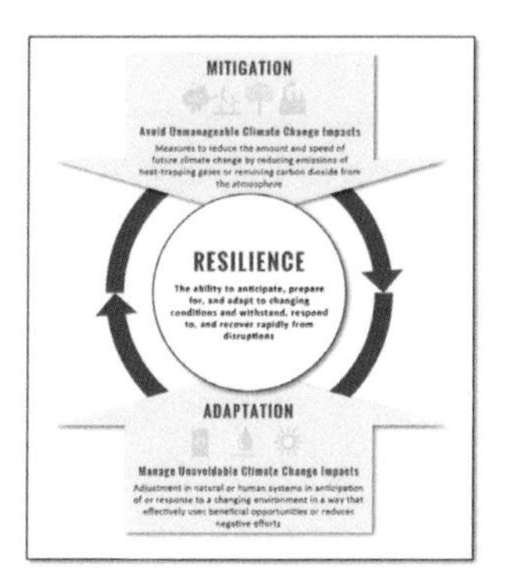

SOURCE: Reproduced from OUSD(A&S), *Department of Defense Climate Adaptation Plan: 2022 Progress Report*, U.S. Department of Defense, October 4, 2022c, cover and p. i; Department of the Navy, Office of the Assistant Secretary of the Navy for Energy, Installations, and Environment, *Climate Action 2030*, May 2022; Department of the Air Force, Office of the Assistant Secretary for Energy, Installations, and Environment, 2022; and Department of the Army, Office of the Assistant Secretary of the Army for Installations, Energy and Environment, *United States Army Climate Strategy*, February 2022.

[9] DoD, 2022b.

[10] OUSD(A&S), 2022c.

[11] John Conger, "And Air Force Makes Three . . . Comparing the U.S. Army, Navy and Air Force Climate Plans," Center for Climate and Security, October 5, 2022.

Each plan presents the services' priorities for addressing climate resilience and each varies in focus (see Table 3.1 for a comparison).[12]

Only the Navy's plan is organized around the LOEs presented in DoD's plan and it describes how these LOEs apply to Navy activities.[13] The Air Force presents goals for each of its three priority areas and a list of key results. For example, for climate-informed decisions, the Air Force is incorporating climate considerations into requirements, acquisition, and supply chain processes. The key results reported include that energy key performance parameters have been incorporated into weapon system capability requirements and that the effects of climate change on supply chains will be identified by FY 2023 to inform risk mitigations by FY 2024.[14] The Army specifically states that Army research, development, test, and evaluation (RDT&E) and modernization investments will enable attainment of the three LOEs. Intermediate objectives for each LOE are presented. For

**TABLE 3.1**

**Comparison of Service Climate Adaptation Plans and Their Structures**

| Service Branch | Focus Areas and Structure of Climate Adaptation Plan |
|---|---|
| Army | • Installations<br>• Acquisition and logistics<br>• Training |
| Navy | • Climate-informed decisionmaking<br>• Train and equip for climate change<br>• Resilient built and natural infrastructure<br>• Supply chain resilience and innovation<br>• Enhance mitigation and adaption through collaboration |
| Air Force | • Maintain air and space dominance in the face of climate change<br>• Make climate-informed decisions<br>• Optimize energy use and pursue alternative energy sources |

SOURCE: Adapted from Conger, 2022.

[12] Conger, 2022.

[13] Department of the Navy, Office of the Assistant Secretary of the Navy for Energy, Installations, and Environment, 2022.

[14] Department of the Air Force, Office of the Assistant Secretary for Energy, Installations, and Environment, 2022.

example, acquisition and logistics has 12 objectives, including "adopt a Buy Clean policy for procurement of construction material with lower embodied carbon emissions, . . . attain net-zero greenhouse gas emissions from all Army procurements by 2050, . . . and implement a revised energy key performance parameter."[15] A common goal across the three service plans is to acquire 100 percent electric nontactical vehicles by 2035.[16]

## DoD Is Focusing on Numerous Research and Development Efforts to Incorporate Environmental Considerations

DoD funds research and development (R&D) activities across the research spectrum, from basic research to new and operational systems development (budget categories 6.1 to 6.8)[17] that correspond to those used by OMB—basic research, applied research, development, R&D facilities, and equipment (although these activities are distributed into DoD RDT&E categories 6.1 to 6.5), as well as other nonexperimental R&D.[18] We identified several DoD activities that promote the development and demonstration of environmental and energy technologies.

DoD has two complementary R&D programs that specifically target environmental and energy technology developments. These programs are SERDP and ESTCP. These are advanced technology development programs (RDT&E budget category 6.3).

---

[15] Department of the Army, Office of the Assistant Secretary of the Army for Installations, Energy and Environment, 2022.

[16] Conger, 2022.

[17] Department of Defense Financial Management Regulation 7000.14-R presents the structure used by the department for R&D funding. The RDT&E budget categories are described in Department of Defense Financial Management Regulation 7000.14-R, Volume 2B, Budget Formulation and Presentation, Chapter 5, September 2022.

[18] John F. Sargent, Jr., *Department of Defense Research, Development, Test, and Evaluation (RDT&E): Appropriations Structure*, Congressional Research Service, R44811, September 7, 2022.

## Strategic Environmental Research and Development Program

Established in 1990 by Congress,[19] SERDP invests in basic research in partnership with the U.S. Department of Energy (DoE) and EPA. SERDP is funded within the Office of the Under Secretary of Defense for Research and Engineering (OUSD[R&E]). Its mission is to contribute to improved mission operations, readiness, and environmental performance while ensuring the health and safety of personnel "by providing new scientific knowledge and developing cost-effective technologies . . . including high-priority requirements," such as addressing per- and polyfluoroalkyl substance (PFAS) contamination, developing fluorine-free fire suppression formulations, and improving corrosion resistance for weapons systems and platforms."[20] There are five program areas: installation energy and water, environmental restoration, munitions response, resource conservation and resilience, and weapon systems and platforms. The weapon system and platforms area within SERDP focuses on developing technologies that reduce emissions and waste during weapon system manufacturing, while reducing current and future liabilities. The program also sponsors a searchable, relational database called ASETSDefense, which provides information and assistance to reduce or eliminate environmental, safety, and occupational health impacts from coatings and treatment processes.[21]

## Environmental Security Technology Certification Program

ESTCP complements SERDP, focusing on enabling the transition of novel technologies to DoD by demonstrating and validating the performance of novel environmental, resilience (including climate resilience), and energy

---

[19] U.S. Code, Title 10, Section 2901, Strategic Environmental Research and Development Program, to Section 2904, Strategic Environmental Research and Development Program Scientific Advisory Board.

[20] OSD, *Department of Defense Fiscal Year (FY) 2023 Budget Estimates: Defense-Wide Justification Book Volume 3 of 5—Research, Development, Test & Evaluation, Defense-Wide*, April 2022, pp. 1–5.

[21] SERDP and ESTCP, homepage, undated-a; and OSD, 2022, pp. 1–5.

technologies to improve readiness, achieve cost savings and efficiencies, and speed remediation of polluted sites on military lands. ESTCP program areas are installation energy and water, weapon systems and platforms, munitions response, environmental restoration, resource conservation and resilience, the DoD Sustainable Technology Evaluation and Demonstration (STED) Program, and installation climate resilience. (Climate resilience funding began in FY 2023.) Examples of technologies pursued in the weapon systems and platforms area include corrosion and repair technologies, sustainable energetics, emissions and waste reduction, and PFAS contamination. ESTCP uses a competitive solicitation process to evaluate proposals from DoD organizations, other federal agencies, and the private sector for program funding.[22] A relatively new element of this program is the STED program, which was developed in response to EO 14057, *Catalyzing Clean Energy Industries and Jobs Through Federal Sustainability.*[23]

## Sustainable Technology Evaluation and Demonstration

Within ESTCP is the STED program, which funds sustainable technology demonstrations to connect manufacturers to potential users at military installations and speed their introduction to the federal marketplace. The program's approach is to identify emerging sustainable technologies, evaluate the technical data against military specifications and other requirements, increase awareness of the environmental and mission benefits of these technologies, and conduct demonstrations at installations and other federal facilities against traditional products. In March 2023, DoD and the General Services Administration (GSA) signed a memorandum of understanding to use product performance and price information gathered by the STED program.[24] Pricing and demand data are provided to the administration's GSA acquisition teams, which establish contracts through GSA's Global Supply Program. Example activities include holding outreach expos with industry at installations to showcase technologies and gather installa-

---

[22] OSD, 2022, pp. 1–7.

[23] Biden, 2021b.

[24] DoD, "DoD, GSA Sign MOU to Bring More Environmental Innovators to Federal Marketplace," press release, March 22, 2023.

tion challenges or needs; hosting a Sustainable Products Center, which is a virtual platform that provides performance and technical information; providing training to DoD personnel on sustainability technologies or products; and serving as a point of contact for questions and success stories.[25] Examples of technologies featured include biobased tires, products with recycled content, and energy-efficient doors and access controls.[26]

Two other activities that promote environmental and energy technologies are the National Defense Center for Energy and Environment (NDCEE) and the Corrosion Control program.

## National Defense Center for Energy and Environment

The Office of the Assistant Secretary of the Army for Installations, Energy, and Environment (Environment, Safety and Occupational Health) is the lead agency for the NDCEE, which is managed by the U.S. Army Environmental Command. The primary objective of the program is to leverage the services' RDT&E investments to facilitate the diffusion of mature technologies (those that are in late-stage development or are commercial off-the-shelf) that address priority environment, safety, occupational health, energy, and climate concerns among the services. Proposals are accepted from the private sector, academia, and government agencies, but these organizations must have a DoD agency identified that will be responsible for transitioning the technology to the warfighter. Proposals should not only address ESOH, energy, and climate priorities but also support readiness, sustainability, and life-cycle cost objectives. Technologies that demonstrate performance in the field are transitioned to the services and other federal agencies.[27]

---

[25] DENIX, "Welcome to the Department of Defense Sustainable Products Center," webpage, undated-e; Erv Koehler, "Ready, Set, STED: Speeding Up Sustainable Acquisition," GSA Blog, December 15, 2022; OSD, 2022, p. 5.

[26] OUSD(A&S), 2022b.

[27] DENIX, "NDCEE Home," webpage, undated-c; NDCEE, *How to Do Business with NDCEE: A Guide for Our Stakeholders*, 2023.

## Corrosion Control

In addition to the specific environmental technology and products programs, DoD supports a corrosion control program that focuses on reducing the environmental, health, and safety effects of equipment and facility maintenance processes. The focus is on maintenance activities, but within this program, funding is provided for the development of safer and environmentally preferred alternatives for coatings, computer-based corrosion prevention design, management, and sustainment training for the acquisition workforce and facilities engineers, as well as technical revisions to corrosion-related military specifications.[28]

## Examples of R&D Activities

In our discussions with SMEs and in the literature review, we identified many examples of potential technologies that might improve resiliency and readiness while addressing such environmental issues as exposure to hazardous chemicals and greenhouse gases (see examples in Figure 3.3).

The electrification of various vehicles and micro-grids for installations are two technologies that have gained some visibility for their potential to improve readiness and resilience while reducing DoD's carbon footprint and greenhouse gases. DoD has been working on hybrid technology for a while, but full electrification of vehicles is gaining traction because of recent EOs. One example is Defense Innovation Unit (DIU) Tactical Hybridization:

> A prototype for a commercial, hybrid-conversion kit for military tactical vehicles has been designed to reduce fuel consumption, improve performance and decrease logistics demand. The Defense Department operates a fleet of more than 250,000 tactical vehicles, which frequently operate in austere conditions. These vehicles often spend as much of their operational time stationary as they do in motion . . . . However, even when stationary, the engines must run in order to power the essential onboard electronics, as well as the heating and cooling systems in the crew compartments. This results in significant fuel consumption while the vehicle idles . . . . [According to the director of

[28] The section of the budget estimate is Program Element 604016D8Z, Department of Defense Corrosion Control Program (OSD, 2022, pp. 4–5).

> the portfolio,] "[b]y integrating an anti-idle capability into our existing fleet of tactical vehicles, the DOD has the opportunity to meaningfully reduce fuel consumption by its operational forces, enabling them to operate longer between refueling . . . . This also promises to reduce the amount of fuel that must be transported into combat zones, reducing the demand on, and risk to, logistics supply chains."[29]

Other examples in Figure 3.3 include such projects as switching to environmentally preferred coatings for airplanes and implementing a hydraulic fluid purification system to reduce the amount and rate of fluid being replaced. Several other projects are led by the Defense Logistics Agency, including a project with the Office of the Assistant Secretary of Defense for Sustainment, and by other stakeholders, who successfully demonstrated the use of a biobased, nontoxic, biodegradable, multipurpose grease to be used for vehicles and equipment, and assigned National Stock Numbers (NSNs) to them so that they can be procured by DoD and civilian federal agencies.[30] Similarly, the Office of the Assistant Secretary of Defense for Sustainment in the Defense Logistics Agency and the Air Force Research Laboratory successfully demonstrated the performance of biobased motor fuel in vehicles and with greater oil change intervals and assigned them new NSNs.[31]

The services also have successful environmentally focused R&D activities. The requirement for Army buildings to achieve a minimum silver level Leadership in Energy and Environmental (LEED) certification has proven successful with the Army Health Clinic in Fort Knox, Kentucky, which achieved a LEED gold certification with the same budget it was provided to achieve a silver certification.[32] The Navy Environmental Sustainability Development to Integration (NESDI) program's work is underway to find a better alternative to cadmium connectors in electronics wiring, which DoD

[29] David Vergun, "Prototype Aims to Reduce Fuel Use, Improve Tactical Vehicle Performance," DoD News, November 24, 2021.

[30] DENIX, "DoD Sustainable Products Center: Biobased Grease Demonstration," webpage, undated-f.

[31] DENIX, "DoD Sustainable Products Center: Biobased Motor Oil Demonstration," webpage, undated-g.

[32] ASA(IE&E), "Sustainability," webpage, updated September 2021.

**FIGURE 3.3**
**Examples of Investment or Procurement Activities Addressing Energy or Environmental Problem Sets**

Environmentally advantaged coatings

Hydraulic fluid purification system

Fully synthetic bio-based motor oils

Bio-based multipurpose greases

High-performance sustainable buildings: Air Force Institute of Technology at Wright-Patterson Air Force Base

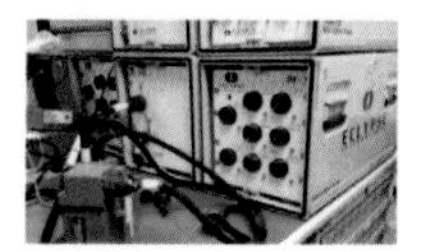

Assessment of cadmium alternatives for connector applications

Remote sensing to detect harmful algal blooms

DIU Tactical Hybridization

SOURCES: Images are respectively from Kelly McNamara, "ENV 101—Introduction to ESOH in Acquisition," briefing slides, U.S. Air Force Life Cycle Management Center, April 25, 2022 (first two images); DENIX, undated-f; DENIX, undated-g; G. Anthonie Riis, June 17, 2017; NAWC-AD Patuxent River Wiring Laboratory (in Navy Environmental Sustainability Development to Integration, *Assessment of Cadmium Alternatives for Connector Applications*, undated); USACE photo by Kansas City District (in Holly Kuzmitski, "Remote Sensing Gives USACE an Edge at Detecting Harmful Algal Blooms," webpage, U.S. Army Corps of Engineers, January 23, 2023); and Army Capt. Joseph Warren, November 24, 2020 (in Vergun, 2021). The appearance of DoD visual information does not imply or constitute DoD endorsement.

has already targeted for elimination because of its toxicity. NESDI's assessment for alternatives is focused on testing a variety of finishes to required performance in real-world settings, a criterion that the alternative connectors have not met so far.[33]

U.S. Army Corps of Engineers districts face the unique task of maintaining hundreds of inland bodies of water and keeping them safe from harmful algal blooms, which are destructive to ecosystems reliant on freshwater and have negative effects on public health, as well as recreation, seafood, and

[33] Navy Environmental Sustainability Development to Integration, undated.

tourism industries. In partnership with government and academic organizations, the Army Corps of Engineers is using remote sensing technologies to collect information about the conditions of the freshwater bodies from a safe distance, then using software to analyze the information to provide early warning of the threat of algal blooms in certain areas.[34] These R&D projects and others demonstrate the variety of technological programs in the department that seek to provide mutual gains for both the DoD mission and the environment.

## DoD Leadership Acknowledges Good Environmental Practices Through Secretary of Defense Environmental Awards

DoD recognizes exemplary environmental stewardship with its Secretary of Defense Environmental Awards, which have been awarded annually to installations, teams, and individuals since 1962.[35] Each of the military services and defense agencies can submit a nomination for each of six installation and three individual/team award categories for achievements in such areas as sustainability, natural resources conservation, cultural resources management, and environmental excellence in weapon system acquisition.[36] Per the *2022 Secretary of Defense Awards: About the Awards* fact sheet, recipients are selected for "outstanding accomplishments in innovative and cost-effective environmental strategies that successfully support mission readiness."[37] Winners receive honors including a letter from the Secretary of Defense and a public display in the Pentagon.[38]

An analysis of winners and nominees listed on the DENIX Secretary of Defense Environmental Awards public page between 2012 and 2022 showed

[34] Kuzmitski, 2023.

[35] DENIX, "Secretary of Defense Environmental Awards Home," webpage, undated-d.

[36] DENIX, undated-d.

[37] DENIX, *2022 Secretary of Defense Environmental Awards: About the Awards*, fact sheet, 2022a.

[38] DENIX, 2022a.

that awardees tend to demonstrate significant achievements in the identification, development, and/or use of safer chemicals (compared with historical status quo or minimum requirements), innovation in ESOH initiatives,[39] and/or exceptional ESOH documentation or ESOH involvement throughout the acquisition process. Table 3.2 shows a sample of six award winners and nominees that we chose to highlight the variety of accomplishments warranting awards or nominations. Successful applicants demonstrated high attention to both technical and logistical detail regarding environmental considerations from regulatory, policy, and personnel dimensions.

---

[39] Environmental considerations are incorporated in the systems engineering process, which refers to *ESOH* rather than *environment*. Applicants tend to follow this convention even when describing primarily environmental activities.

**TABLE 3.2**

## Examples of Secretary of Defense Environmental Awards

| Recipient | Year | Accomplishments | Winner or Nominee | Award Type |
|---|---|---|---|---|
| C-130 Program Office and Support Team, Robins Air Force Base, Georgia | 2022 | After a multi-year process, identified three non-chromate corrosion-inhibiting sealants, eliminating 13,500 pounds of hazardous chromate waste—protecting personnel while saving more than $250,000 per year | Winner | Environmental Excellence in Weapon Systems Acquisition |
| Naval Supply Systems Command, Weapons Systems Support | 2021 | Used information from EPA's Safer Choice program to compile a database of environmentally friendly chemical products for purchase by Navy bases. These products are pre-approved for purchase, fast-tracking the process and increasing capabilities. Team also published the standard operating procedures and technical guidance for the system | Winner | Sustainability |
| Combat Rescue Helicopter Program ESOH Team | 2018 | Early and extraordinary ESOH involvement in the acquisition process led to significant impacts on the reduction of HAZMAT usage on the CRH. This includes the elimination of hexavalent chromium paints from the exterior and interior structural surfaces of the aircraft and eliminating 40 percent of the HAZMAT across the airframe, avionics, and maintenance technical documentation[a] | Winner | Environmental Excellence in Weapon Systems Acquisition |
| KC-46A Program ESOH Team, Wright Patterson AFB | 2016 | Achievements include eliminating use of halon and reduced use of hexavalent chromium by integrating deeply into systems engineering planning and execution throughout acquisition process, beyond what was expected or required[b] | Winner | Environmental Excellence in Weapon Systems Acquisition |

## Table 3.2—Continued

| Recipient | Year | Accomplishments | Winner or Nominee | Award Type |
|---|---|---|---|---|
| Joint Light Tactical Vehicle ESOH Working Group, Michigan | 2016 | Comprised of a myriad of SMEs and stakeholders, excelled in coordination across functions to ensure ESOH considerations were made throughout acquisition process, crafting key requirements before Milestone C | Nominee | Environmental Excellence in Weapon Systems Acquisition |
| Fairchild AFB Environmental Management System Cross-Functional Team, Spokane, Washington | 2015 | Increased use of environmentally preferable products by labeling GPP items in base supply store, and trained all contracting personnel on green procurement, including monthly trainings for GPC holders | Nominee | Environmental Excellence in Weapon Systems Acquisition |

SOURCES: DENIX, *2022 Secretary of Defense Environmental Awards: Environmental Excellence in Weapon Systems Acquisition, Team—C-130 Program Office and Support Team*, 2022b; DENIX, *2021 Secretary of Defense Environmental Awards: Sustainability, Individual/Team—Naval Supply Systems Command, Weapon Systems Support, Pennsylvania*, 2021a; DENIX, *2018 Secretary of Defense Environmental Awards: Environmental Excellence in Weapon Systems Acquisition, Team—Combat Rescue Helicopter Program ESOH Team*, 2018; DENIX, *2016 Secretary of Defense Environmental Awards: Environmental Excellence in Weapon Systems Acquisition, Large Program—KC-46A Program Environment, Safety, and Occupational Health Team*, 2016b; DENIX, *2016 Secretary of Defense Environmental Awards: Environmental Excellence in Weapon Systems Acquisition, Large Program—Joint Light Tactical Vehicle Environmental, Safety, and Occupational Health Working Group, Michigan*, 2016a; DENIX, *2015 Secretary of Defense Environmental Awards: Environmental Excellence in Weapon Systems Acquisition, Small Team—Fairchild Air Force Base*, 2015.

NOTES: AFB = air force base. CRH = combat rescue helicopter. GPC = government purchase card. GPP = Green Procurement Program. HAZMAT = hazardous materials.

[a] "The CRH Program's decision to mitigate risks by eliminating proven legacy Cr6+ coatings on both external and internal surfaces of the aircraft could increase risk of internal, and more difficult to detect, corrosion. This represented a major departure from the program's legacy system-based acquisition strategy, but it was the right thing to do" (DENIX, 2018, p. 4).

[b] "This exemplary ESOH effort was accomplished on a commercial derivative system acquisition program, an acquisition approach that often discourages additional ESOH 'improvements' beyond the baseline established by the underlying commercial system" (DENIX, 2016, p. 2).

CHAPTER 4

# Acquisition and Environmental Policy and Process

In this chapter, we describe the acquisition and environmental policies and processes the DoD acquisition workforce uses to incorporate environmental considerations into acquisition decisionmaking. These policies and processes enable DoD to address cost (including life-cycle costs), resource, and energy efficiency and resiliency in acquisition planning and practice, including incorporating environmental performance in source selection.

## The DoD Acquisition Policy Environment Distinguishes Between Weapon Systems and Goods and Services

DoD implements statutes and EOs and incorporates environmental considerations into acquisition decisions primarily in compliance with the FAR and requirements that are embedded in some of the formal Adaptive Acquisition Framework Pathways.[1] These policy mechanisms generally

[1] As mentioned previously, DoD policy for the defense acquisition system seems to emphasize ESOH risks and requirements management. ESOH is

> the combination of disciplines that encompass the processes and approaches for addressing laws, regulations, Executive Orders (EO), DoD policies, environmental compliance, and hazards associated with environmental impacts, system safety (e.g., platforms, systems, system-of-systems, weapons, explosives, software, ordnance, combat systems), occupational safety and health, hazardous materials management, and pollution prevention. (Military Standard 882E, 2012)

It is implemented through system safety engineering.

distinguish between the acquisition of goods and services, but DoD established the DoD Sustainable Procurement Program (DSPP) Working Group (DSPPWG) to implement its DSPP, which applies more generally to both goods and services.[2] The DSPP framework includes policy, planning, implementation, evaluation, and corrective action. The overarching policy is to "give preference to procurement of sustainable goods and services that use or supply sustainable goods" unless a statutory exception applies or unless DoD cannot acquire a good or service that meets performance requirements, meets performance schedule, or is obtainable at a reasonable price.[3]

## The FAR's Environmental Requirements

FAR Part 23 provides general environmental guidance for all contract actions; project and program managers have discretion in navigating the process and applying that guidance as appropriate for their project or program. FAR Part 23 covers environment, energy and water efficiency, renewable energy technologies, occupational safety, and drug-free workplace and FAR supplements; DoD, service, and component policy; and guidance. FAR 23.103, Sustainable Acquisitions, requires that 95 percent of new contract actions for the supply of products and acquisition of services (including construction) be for products that are "Energy-efficient (ENERGY STAR® or Federal Energy Management Program [FEMP]-designated); Water-efficient; Biobased; Environmentally preferable (e.g., EPEAT®-registered, or non-toxic or less toxic alternatives); Non-ozone depleting; or Made with recovered materials."[4]

## Environmental Requirements Within the Adaptive Acquisition Framework

OUSD(A&S) developed the Adaptive Acquisition Framework (AAF), which consists of six acquisition pathways for different kinds of systems: Urgent Capability Acquisition, Middle Tier of Acquisition, Major Capability Acqui-

2 DoDI 4105.72, 2018.

3 DoDI 4105.72, 2018.

4 FAR 23.103, 2023b.

sition, Software Acquisition, Defense Business Systems, and Acquisition of Services.

Of the six acquisition pathway policies, only the DoDIs that provide guidance for urgent capability acquisition and major capability acquisition include environmental considerations (see Table 4.1). Engineering and test and evaluation are the only two functional policies that include environmental considerations on the AAF website.[5]

There is no functional policy for environmental considerations, similar to other cross-cutting functional areas like cybersecurity or test and evalu-

**TABLE 4.1**
**Tabulation of Acquisition-Related DoDIs with Environmental Considerations**

| Policy Number | Title and Date | Environmental Considerations |
|---|---|---|
| Department of Defense Directive (DoDD) 5000.01 | *The Defense Acquisition System*, July 28, 2022 | Yes |
| DoDI 5000.02 | *Operation of the Adaptive Acquisition Framework*, June 8, 2022 | No |
| DoDI 5000.71 | *Rapid Fulfillment of Combatant Commander Urgent Operational Needs and Other Quick Action Requirements*, October 18, 2022 | No |
| DoDI 5000.73 | *Cost Analysis Guidance and Procedures*, March 13, 2020 | No |
| DoDI 5000.74 | *Defense Acquisition of Services*, June 24, 2021 | No |
| DoDI 5000.75 | *Business Systems Requirements and Acquisition*, January 24, 2020 | No |
| DoDI 5000.80 | *Operation of the Middle Tier of Acquisition (MTA)*, December 30, 2019 | No |
| DoDI 5000.81 | *Urgent Capability Acquisition*, December 31, 2019 | Yes |
| DoDI 5000.82 | *Acquisition of Information Technology (IT)*, April 21, 2020 | No |

[5] DAU, "Adaptive Acquisition Framework Pathways: Acquisition Policies," webpage, undated-b.

**Table 4.1—Continued**

| Policy Number | Title and Date | Environmental Considerations |
|---|---|---|
| DoDI 5000.83 | *Technology and Program Protection to Maintain Technological Advantage*, May 21, 2021 | No |
| DoDI 5000.84 | *Analysis of Alternatives*, August 4, 2020 | No |
| DoDI 5000.85 | *Major Capability Acquisition*, November 4, 2021 | Yes |
| DoDI 5000.86 | *Acquisition Intelligence*, September 11, 2020 | No |
| DoDI 5000.87 | *Operation of the Software Acquisition Pathway*, October 2, 2020 | No |
| DoDI 5000.88 | *Engineering of Defense Systems*, November 18, 2020 | Yes |
| DoDI 5000.89 | *Test and Evaluation*, November 19, 2020 | Yes |
| DoDI 5000.90 | *Cybersecurity for Acquisition Decision Authorities and Program Managers*, December 31, 2020 | No |
| DoDI 5000.91 | *Product Support Management for the Adaptive Acquisition Framework*, November 4, 2021 | No |
| DoDI 5000.95 | *Human Systems Integration in Defense Acquisition*, April 1, 2022 | No |
| DoDI 5010.44 | *Intellectual Property (IP) Acquisition and Licensing*, October 16, 2019 | No |

ation. In addition to DoDIs, some major DoD-level acquisition guides and manuals address environmental considerations (see Table 4.2). However, there is no guidebook focused on environmental considerations in acquisition. The lack of a functional policy or guidebook for environmental considerations could be considered a gap in knowledge and tools.

Several interviewees argued that the AAF pathway policies are not the appropriate place to prescribe or issue environmental compliance or guidance and that DoD uses functional policies, service and agency acquisition policies, and functional guidebooks to cover environmental considerations, although guidebooks are not mandatory (see Table 4.3).[6] Service and agency acquisition policy and functional guidance related to engineering, systems engineering, product support management, human systems integration, and test and evaluation include environmental life-cycle considerations related to ESOH, and these are incorporated early in the process (before

**TABLE 4.2**

**Other DoD-Level Major Acquisition Guidebooks and Manuals**

| Title and Date | Environmental Considerations |
|---|---|
| *DoD Cybersecurity Test and Evaluation Guidebook*, February 10, 2020 | No |
| *Analysis of Alternatives Cost Estimating Handbook*, July 2021 | Yes |
| *Intelligence Support to the Adaptive Acquisition Framework (ISTAAF) Guidebook*, September 2021 | No |
| *Engineering of Defense Systems Guidebook*, February 2022 | Yes |
| *Requirements for the Acquisition of Digital Capabilities Guidebook*, February 2022 | No |
| *Systems Engineering Guidebook*, February 2022 | Yes |
| *Guide to DoD International Acquisition and Exportability Practices*, March 2022 | No |
| *Human Systems Integration Guidebook*, May 2022 | Yes |
| *Product Support Manager Guidebook*, May 2022 | Yes |
| *Department of Defense Technology and Program Protection Guidebook*, July 2022 | No |
| *A Guide to Program Management Knowledge, Skills, and Practices*, August 1, 2022 | No |
| *A Guide to Program Management Business Processes*, August 4, 2022 | No |
| *Source Selection Procedures*, August 20, 2022 | No |
| *Test and Evaluation Enterprise Guidebook*, August 2022 | Yes |

[6] Discussion with subject-matter experts, January 2023.

**TABLE 4.3**

**Sample of Service-Level Acquisition and Functional Policies and Guidebooks**

| Policy Number | Title | Environmental Considerations |
|---|---|---|
| Air Force Instruction 63-101 | *Integrated Life Cycle Management*, November 23, 2021 | Yes |
| Department of the Army Pamphlet 70-3 | *Army Acquisition Procedures*, September 17, 2018 | Yes |
| Secretary of the Navy Instruction 5000.2G | *Department of the Navy Implementation of the Defense Acquisition System and the Adaptive Acquisition Framework*, April 8, 2022 | Yes |

Milestone B).[7] Acquisition decisions and documentation that address environmental considerations include the following:

- Acquisition Strategy
- Assessment of Technical Risk and Development of Mitigation Plans
- Contract Data Requirements Lists (CDRLs)
- Cost Analysis Requirements Description (CARD)
- Design Considerations
- ESOH planning
- HAZMAT Management
- Human Systems Integration (HSI) Plan
- Life Cycle Management Plan (LCMP)
- Life Cycle Sustainment Plan (LCSP)

[7] OUSD(R&E), Office of the Deputy Director for Engineering, *Engineering of Defense Systems Guidebook*, U.S. Department of Defense, February 2022a; OUSD(R&E), Office of the Deputy Director for Engineering, *Systems Engineering Guidebook*, U.S. Department of Defense, February 2022b; Office of the Deputy Assistant Secretary of Defense for Product Support, *Product Support Manager Guidebook*, U.S. Department of Defense, May 2011, updated May 24, 2022; OUSD(R&E), Office of the Deputy Director for Engineering, *Human Systems Integration Guidebook*, U.S. Department of Defense, May 2022c; OUSD(R&E) and Office of the Director for Operational Test and Evaluation, *Test and Evaluation Enterprise Guidebook*, U.S. Department of Defense, August 2022.

- Manufacturing readiness
- NEPA/EO 12114 Compliance Schedule
- Product Support Strategy (PSS)
- PESHE
- System Performance Specification
- System Safety Engineering program and management planning, preliminary hazard analysis, Hazard Tracking System (HTS)
- Systems Engineering Management Plan (SEMP)
- Systems Engineering Plan (SEP)
- Technical Baseline Documentation/Digital Artifacts
- Technical Risk Assessment
- Test and Evaluation Master Plan (TEMP).

Several planning activities and documents address environmental considerations directly, including ESOH planning, HAZMAT management, NEPA/EO 12114, and the PESHE. The other activities and associated documents address aspects of environmental issues, planning, and compliance as part of broader functional topics; for example, the Acquisition Strategy and SEP.

In addition to acquisition policies and required documentation that address environmental considerations, DoD has issued numerous policies that are directly related to the environment, many of which govern acquisition activities. Table 4.4 lists a sample of these policies and a brief description of each.

## Incorporating Environmental Considerations in the Weapon System Acquisition Life Cycle

DoD environmental policy and guidance frame improvements in energy efficiency and environmental compliance in terms of improvements in mission outcomes, capabilities, resilience, and readiness, as well as in terms of reductions in life cycle costs. There are points throughout the weapon system life cycle at which discussions of environmental considerations can happen. For environmental efficiencies to be "baked in" throughout the life cycle of the program, they need to be "designed in" earlier in the process.

**TABLE 4.4**

## Sample of DoD's Environmental Policies

| Policy Number | Title | Applicability |
|---|---|---|
| DoDD 4715.1E | *Environment, Safety, and Occupational Health (ESOH)*, December 30, 2019 | DoD components<br>• Operations, activities, and installations worldwide |
| DoDD 4715.11 | *Environmental and Explosives Safety Management on Operational Ranges Within the United States*, August 31, 2018 | DoD components<br>• All operational ranges in the United States |
| DoDD 4715.12 | *Environmental and Explosives Safety Management on Operational Ranges Outside the United States*, August 31, 2018 | DoD components<br>• All operational ranges outside the United States |
| DoDI 4105.72 | *Procurement of Sustainable Goods and Services*, August 31, 2018 | DoD components<br>• Does not apply to weapon systems, nonappropriated fund instrumentalities and procurement, and alternative fuels for operational platform purposes |
| DoDI 4140.25 | *DoD Management Policy for Energy Commodities and Related Services*, December 30, 2019 | DoD components |
| DoDI 4170.11 | *Installation Energy Management*, August 31, 2018 | DoD components<br>• All activities that affect the supply, reliability, and consumption of facility energy |
| DoDI 4715.02 | *Regional Environmental Coordination*, August 31, 2018 | DoD components<br>• Actions by DoD components outside the United States on installations under DoD control "support functions for U.S. military vessels, ships, aircraft, or space vehicles provided by the DoD Components" (DoDI 4715.05, 2018, p. 1) |

**Table 4.4—Continued**

| Policy Number | Title | Applicability |
|---|---|---|
| DoDI 4715.05 | *Environmental Compliance at Installations Outside the United States*, August 31, 2018 | DoD components<br>• Actions by DoD components outside the United States on installations under DoD control "support functions for U.S. military vessels, ships, aircraft, or space vehicles provided by the DoD Components" (p. 1) |
| DoDI 4715.06 | *Environmental Compliance in the United States*, August 31, 2018 | DoD components<br>• Operations, activities, and installations in the United States |
| DoDI 4715.07 | *Defense Environmental Restoration Program (DERP)*, August 31, 2018 | DoD components<br>• Environmental restoration within the United States |
| DoDI 4715.15 | *Environmental Quality Systems*, July 8, 2019 | DoD components<br>• "[A]ctivities and programs involving the collection, management, and use of environmental data, supporting all applicable environmental laws and regulations, at DoD operations, activities, and installations worldwide, including government-owned/contractor-operated facilities and formerly used defense sites" (p. 3) |
| DoDI 4715.17 | *Environmental Management Systems*, August 31, 2018 | DoD components<br>• All DoD facilities and/or organizations worldwide |
| DoDI 4715.18 | *Emerging Chemicals (ECs) of Environmental Concern*, September 4, 2019 | DoD components<br>• Operations, activities, and installations worldwide |
| DoDI 4715.21 | *Climate Change Adaptation and Resilience*, August 31, 2018 | DoD components<br>• "All DoD operations worldwide unless superseded by international agreement" (p. 3) |

**Table 4.4—Continued**

| Policy Number | Title | Applicability |
|---|---|---|
| DoDI 4715.22 | *Environmental Management Policy for Contingency Locations*, August 31, 2018 | DoD components<br>• Contingency locations, all phases in the life cycle of contingency locations, and training areas associated with transition or closure requirements for contingency locations |
| DoDI 4715.24 | *The Readiness and Environmental Protection Integration (REPI) Program and Encroachment Management*, March 27, 2019 | DoD components<br>• Military installations within the United States |
| DoDI 6055.05 | *Occupational and Environmental Health (OEH)*, August 31, 2018 | DoD components<br>• Worldwide, although statutory requirements in this DoDI generally apply only within the United States<br>• Contractor operations and personnel deployed to contingency locations |
| DoDI 6055.20 | *Assessment of Significant Long-Term Health Risks from Past Environmental Exposures on Military Installations*, June 10, 2019 | DoD components<br>• Military installations |

DoD infuses environmental considerations through the system engineering process, design interface, and product support processes.

For example, a KPP exists for energy efficiency that needs to be considered during requirements development. This is a statutorily mandated energy KPP.[8] Through the Climate Working Group, DoD "assessed the extent to which it applied the energy KPP over 44 joint programs, and found inconsistent application of the energy KPP and an uneven prioritization of energy supportability across joint programs."[9] The OSD and Army climate

[8] U.S. Code, Title 10, Section 2911.

[9] Kathleen H. Hicks, Deputy Secretary of Defense, "Energy Supportability and Demand Reduction in Capability Development," memorandum for Secretaries of the

strategies make reference to the energy KPP as driving the department to include energy use in the requirements process, which enables energy performance or efficiency to be considered in source selection decisions.[10]

DoD has had mixed success with applying environmental considerations early in acquisition programs for a variety of reasons. According to SMEs, discussions with stakeholders are more likely to happen closer to Milestone B or when a potential environmental situation is on the horizon.

Figure 4.1 provides the generalized life cycle of a weapon system. The figure identifies the following points throughout the process that are opportunities to consider or to take action on regarding environmental considerations:

- The Materiel Solution Analysis Phase includes consideration of the types of technologies that may be employed to establish a potential capability, should be evaluated during the Analysis of Alternatives, and may inform materiel requirements development. At a minimum, a potential hazard list and ESOH criteria should be provided during those discussions.
- While conducting source selection in the Technology Maturation and Risk Reduction Phase, ESOH considerations should be evaluated if they were included in the Request for Proposal (RFP).
- During the Engineering and Manufacturing Development Phase, detailed system design and engineering take place and the decisions concerning test and evaluation (T&E) are documented in the T&E Master Plan. This is another point in the life cycle at which an ESOH analysis or risk assessment can be conducted; HAZMAT can be identified along with ESOH requirements in technical publications. NEPA/EO 12114 analyses should also be conducted at this time. An ESOH

Military Departments; Chairman of the Joint Chiefs of Staff; and Under Secretary of Defense for Acquisition and Sustainment, April 21, 2022, p. 1.

[10] OUSD(A&S), 2022c; Department of the Army, Office of the Assistant Secretary of the Army for Installations, Energy and Environment, 2022.

FIGURE 4.1
**Incorporating Environmental Considerations into Weapon System Acquisition**

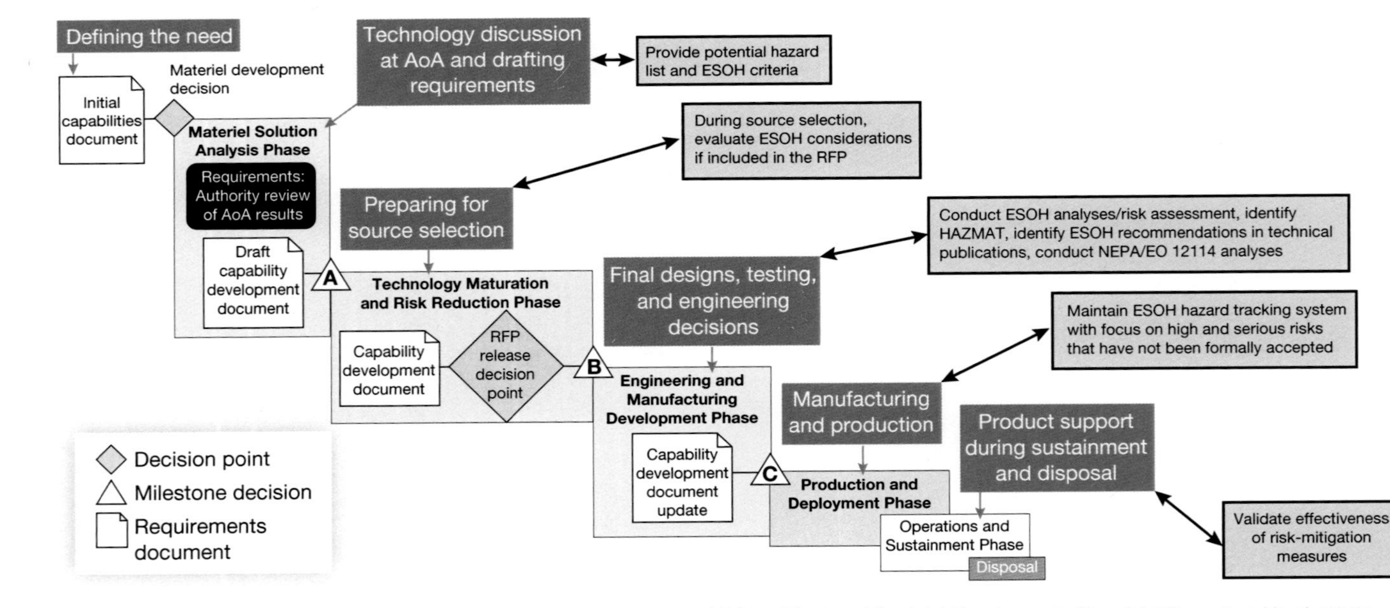


SOURCE: Adapted from Chairman of the Joint Chiefs of Staff Instruction 5123.011, *Charter of the Joint Requirements Oversight Council and Implementation of the Joint Capabilities Integration and Development System*, October 30, 2021, p. D-6. Features information from Office of the Deputy Under Secretary of Defense for Installations and Environment and Office of the Deputy Under Secretary of Defense for Acquisition and Technology, *Environment, Safety, and Occupational Health (ESOH) in Acquisition: Integrating ESOH into Systems Engineering*, April 2009.

NOTE: AoA = Analysis of Alternatives. RFP = request for proposal.

hazard tracking system that is focused on high and serious risks that have not been formally accepted should be maintained.[11]

- During the Operations and Sustainment Phase, product support provides an opportunity to validate the effectiveness of risk-mitigation measures that were put in place earlier in the life cycle.

## Incorporating Environmental Considerations in the Acquisition of Services Life Cycle

The AAF provides a seven-step process outlining the acquisition of services process within DoD, as specified in DoDI 5000.74 and the Defense Acquisition University (DAU) Service Acquisition Mall.[12] The steps are nested within Plan-Develop-Execute groupings, containing Steps 1–3, 4–5, and 6–7, respectively.[13] Figure 4.2 outlines these steps. Within this process, there are several points at which environmental considerations can be discussed and prioritized:[14]

- During Steps 1 and 2 of the Plan Phase, an analysis of the potential risks, opportunities, and gaps in requirements can be conducted, along with an analysis of compliance with current environmental statutes, EOs, regulations, and policy. While market research is being conducted in Step 3 of the Plan Phase, DoD project managers and contracting personnel can assess environmental considerations by requesting environmentally relevant information from service providers with respect

[11] Per DoDI 5000.88, *Engineering of Defense Systems*, U.S. Department of Defense, November 18, 2020.paragraph 3.6.e.(1)(b)1; and Military Standard 882E, 2012.

[12] DAU, "Adaptive Acquisition Framework Pathways: Acquisition of Services," webpage, undated-c; DAU Service Acquisition Mall, "Service Acquisition Steps," webpage, undated.

[13] DAU, undated-c.

[14] FAR Part 23 provides general environmental guidance for all contract actions, but project managers have discretion in navigating the process, as appropriate for their project (Federal Acquisition Regulation, Part 23, Environment, Energy and Water Efficiency, Renewable Energy Technologies, Occupational Safety, and Drug-Free Workplace, March 16, 2023a).

FIGURE 4.2
**Incorporating Environmental Considerations into the Acquisition of Services Process**

| Plan | | | Develop | | Execute | |
|---|---|---|---|---|---|---|
| Step 1: Form the team | Step 2: Review current strategy | Step 3: Perform market research | Step 4: Define requirements | Step 5: Develop acquisition strategy | Step 6: Execute strategy | Step 7: Manage performance |
| | Analyze gaps in compliance with current regulations | Search for vendors who meet or exceed regulatory requirements | PM and CO work with user to define additional environmental requirements as applicable and ensure that compliance will be achieved | Design solicitation and evaluation factors to prioritize environmental performance requirements | | Ensure regulatory compliance throughout contract period |

SOURCE: Adapted from DAU, undated-c.
NOTE: PM = project manager. CO = contracting officer.

to the vendor's capabilities, any recent innovations, best practices, industry trends, and alternative ways of meeting the agency's need to better inform their assessment of the landscape of vendors who can meet or exceed requirements.

- During the Develop Phase, the project manager and contracting officer can work with the user to define environmental requirements, performance objectives, performance standards, and methods of inspection, as applicable during Step 4. While developing the acquisition strategy in Step 5, the government can design solicitation and source-selection evaluation factors, subfactors, or criteria that prioritize environmental performance requirements and considerations and that discriminate between offerers on these criteria.
- Finally, during the Execute Phase (Steps 6 and 7), the project manager and contracting officer can monitor performance and compliance with environmental policy and other contract requirements.[15] The result and output of monitoring activities should include recur-

[15] DoD can terminate contracts with vendors who fail to uphold regulatory requirements (e.g., wastewater treatment plants exhibiting poor waste management practices)

ring reviews with the service providers and other stakeholders, along with the annual submission of Contractor Performance Assessment Reports (CPARs).

## Incorporating Environmental Considerations in the Acquisition of Commodities and Supplies Life Cycle

Whereas the acquisition of services is governed by the AAF services pathway, commodity procurement does not have a unique AAF pathway, although it is still governed by the FAR and applicable DoD component policy. FAR Part 23 specifies environmental requirements for commodity procurement—for example, by requiring electronics to be Energy Star certified, when possible.[16]

The study team identified a five-step process for commodity acquisition: Profile Commodity, Conduct Supply Market Analysis, Develop Commodity Strategy, Issue RFx (i.e., a solicitation, such as a Request for Quote or Request for Proposal) and Negotiate, and Implement and Manage Performance.[17] As with the acquisition of services, the team next identified how environmental considerations could be discussed and prioritized within each step. From this high-level perspective, the process and potential inputs closely resemble those identified for the acquisition of services (see Figure 4.3):

- During the Profile Commodity phase, the government defines its requirements, including those related to environmental considerations; different methods of meeting the agency's need; detailed specifications; and total spend.

(DENIX, *2021 Secretary of Defense Environmental Awards: Sustainability, Individual/Team—Whiteman AFB Environmental Element Sustainability Team*, 2021b).

[16] FAR Part 23, 2023a.

[17] Rene G. Rendon, *Commodity Sourcing Strategies: Supply Management in Action*, Naval Postgraduate School, January 31, 2005, p. 24. This five-step process was adapted from Figure 4, which depicts a six-part commodity strategic sourcing process.

**FIGURE 4.3**
**Incorporating Environmental Considerations into Commodities and Supplies Acquisition**

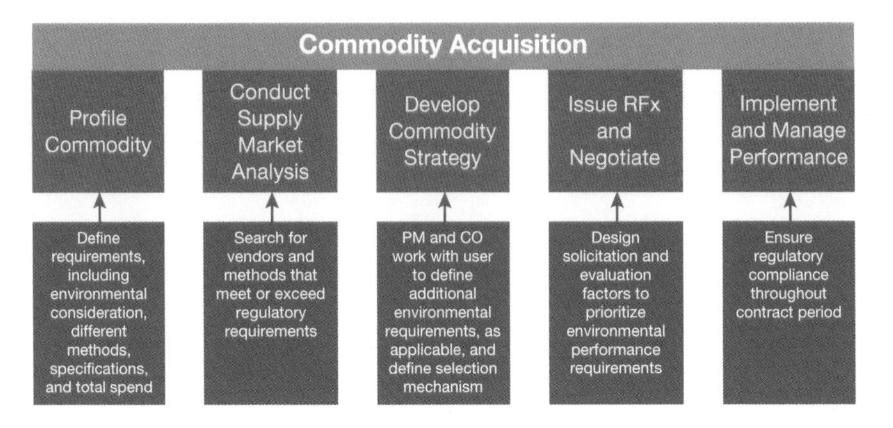


SOURCE: Adapted from Rendon, 2005, p. 24.
NOTE: PM = project manager. CO = contracting officer.

- During the Conduct Supply Market Analysis phase, government personnel can request information from industry on trends and ways to meet the government's needs, as well as search for vendors and methods of achieving capability requirement that meet or exceed environmental requirements.
- During the Develop Commodity Strategy phase, the project manager and contracting officer work with the end user or customer to define additional environmental requirements, as applicable, for the commodity or supply and define the selection mechanism.
- During the Issue RFx and Negotiate phase, the government designs the solicitation and evaluation factors, subfactors, or criteria to prioritize environmental performance requirements and considerations that allow the government to discriminate between offerers.
- During the Implement and Manage Performance phase, the government ensures that the contractor complies with all environmental requirements throughout the contract period.

In mapping the commodity procurement process, we noted that, although there are distinct phases or decision points at which issues associated with environmental management, compliance, and impact could be addressed, there does not appear to be an organization or a functional position that can act as an environmental adviser to the project manager and contracting officer. This observation applies to the acquisition of services as well. This contrasts with the weapon system acquisition process, in which an organization within the services' life-cycle management or system commands has the subject-matter expertise to provide advice. It is likely that the energy, installation, and environment organizations within the services and OSD contain SMEs that could provide that environmental advice for the procurement of goods and services, but they might not have the capacity. The large number of commodity and acquisition of services procurement actions suggests that having environmental SMEs closer to those processes would be more impactful. The knowledge and tools to support this activity exist, but having a designated functional position (similar to the small business adviser that many contracting offices have) integrated into the contracting process might help foster making more–environmentally preferred decisions.

## Incorporating Environmental Considerations in the Military Construction Life Cycle

The policy of the federal government is to implement "high-performance sustainable building design, construction, renovation, repair, commissioning, operation and maintenance, management, and deconstruction practices"[18] so that agencies comply with the Guiding Principles for Federal Leadership in High-Performance and Sustainable Buildings. The six guiding principles are as follows:

- employ integrated design principles
- optimize energy performance
- protect and conserve water
- enhance the indoor environmental quality

[18] FAR, Part 36, Construction and Architect-Engineer Contracts, Subpart 36.104, Policy, February 13, 2023c.

- reduce the environmental impact of materials
- assess and consider building resilience.[19]

The study team also identified a five-step, high-level process developed for DoD construction acquisition (Figure 4.4): Planning, Design, Solicitation Preparation and Evaluation, Construction and Manage Performance, and Operation and Maintenance (O&M) and Facility Maintenance and Reinvestment.[20] The team identified the following opportunities for environmental considerations within the construction acquisition life cycle:

- During the Planning phase, DoD components must perform an Environmental Impact Analysis, including receiving permits and identifying environmental compliance requirements.
- During the Design phase, the government should ensure that the contractor's design complies with DoD standards, including the Unified Facilities Criteria and High-Performance and Sustainable Building requirements.
- During the Solicitation Preparation and Evaluation phase, the government can design the solicitation and evaluation factors, subfactors, or criteria to prioritize environmental performance requirements and decide among offerers.
- During the Construction and Management of Performance phase, the government can emphasize environmental requirements and considerations during the construction kickoff meeting and continue throughout construction performance in recurring meetings. Additionally, the contracting officer's representative should ensure statutory, regulatory, and contract compliance throughout construction during inspections; notify the contracting officer of any issues; and ensure that required documentation is completed.
- During the O&M and Facility Maintenance and Reinvestment phase, DoD components can identify opportunities during sustainment, res-

---

[19] EPA, "Guiding Principles for Sustainable Federal Buildings," webpage, May 4, 2022.

[20] Constantine Samaras, Abigail Haddad, Clifford A. Grammich, and Katharine Watkins Webb, *Obtaining Life-Cycle Cost-Effective Facilities in the Department of Defense*, RAND Corporation, RR-169-OSD, 2013.

toration, and modernization to incorporate materials and technologies into the structure that better comply with the Guiding Principles for Federal Leadership in High-Performance and Sustainable Buildings and implement them, as appropriate.

**FIGURE 4.4**
**Incorporating Environmental Considerations into Construction Life Cycle**

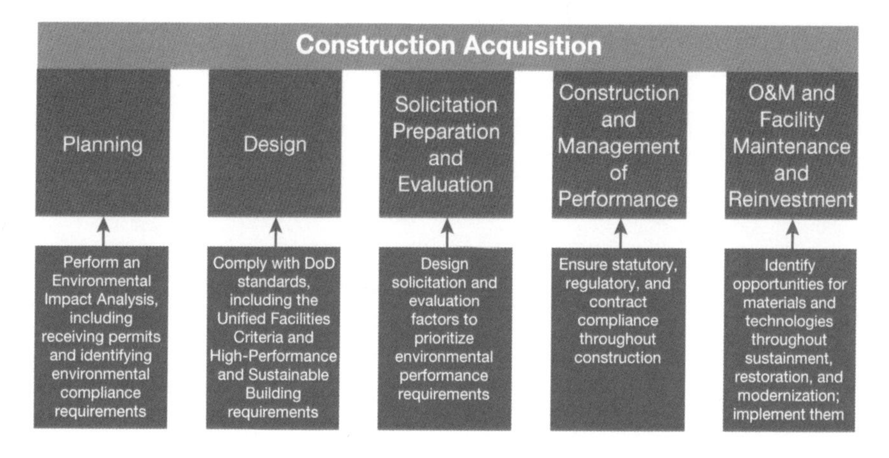


SOURCE: Adapted from Figure S.1 in Samaras, et al., 2013, p. xii.

CHAPTER 5

# Education, Training, and Other Resources

## Education and Training for the Acquisition Workforce on Environmental Considerations

From a DoD-wide perspective, leadership is working toward educating, training, and engaging the DoD workforce about climate change through the CLSWG; however, the DoD acquisition workforce also has broader considerations involving the environment to consider. In this chapter, we discuss how the acquisition workforce currently acquires education and training on environmental considerations.

Of the six acquisition functional areas within the Defense Acquisition System (business financial management and business cost estimating, contracting, engineering and technical management, life cycle logistics, program management, and test and evaluation), none focuses specifically on environmental considerations.[1] However, environmental considerations are relevant to each functional area. For example, engineers designing a weapon system must understand energy consumption or how to design efficiencies into the system. During fielding, logisticians would need to consider the potential environmental repercussions of using certain chemicals for decades. The entire responsibility of incorporating environmental considerations does not neatly fall under the purview of one functional area. Therefore, educational opportunities and training that address environmental policy, requirements, and procedures must be provided to personnel

[1] DAU, "Acquisition Functional Areas," webpage, undated-a.

across all functional areas. In fact, several institutions develop and provide educational opportunities and training on environmental considerations to DoD's acquisition professionals. These institutions include government educational organizations, public and private colleges and universities, and commercial-sector partnerships. Notable examples of government institutions include the DAU,[2] the Air Force Institute of Technology (AFIT) and its Department of Systems Engineering and Management,[3] the Naval Postgraduate School (NPS) and its acquisition management degree programs,[4] and the Federal Acquisition Institute.[5] Some courses (such as DAU's Sustainable Procurement Program course or AFIT's Environmental Restoration Program Management course) focus on environmental considerations, and some incorporate them as one facet of a broader acquisition topic (such as NPS's Principles of Acquisition and Contract Management course). Generally, most courses focus on environmental considerations in the context of a broader topic, such as program management or contracting, but specific degrees or focus areas, such as civil engineering, tend to have courses focused on environmental considerations. See Figure 5.1 for some options of courses available at these sample institutions.

Along with courses provided by outside institutions and those in partnership with DoD or the services, specific offices that are focused on environmental considerations within DoD provide information as needed, such as ENV 101: Introduction to ESOH in Acquisition, which is provided by the Air Force Life Cycle Management Center Acquisition Environmental Integration Office.[6] A variety of courses are provided to support the acquisi-

---

[2] DAU, "iCatalog Home Page," webpage, undated-d.

[3] AFIT, Graduate School of Engineering and Management, *Graduate School Academic Catalog 2021–2022*, undated.

[4] Naval Postgraduate School, "Department of Defense Management," webpage, undated.

[5] Federal Acquisition Institute, homepage, undated.

[6] Air Force Life Cycle Management Center, "AFLCMC Focus Week Course Catalog, 25–29 April 2022," briefing slides, March 29, 2022.

**FIGURE 5.1**
## Sample Institutions and Courses Available to the Acquisition Workforce

| Air Force Institute of Technology | Naval Postgraduate School | Defense Acquisition University | Federal Acquisition Institute |
|---|---|---|---|
| Intro to Environmental Restoration Program: WENV 021 | MBA in Acquisition Contract Management: 815 (degree) | DoD Sustainable Procurement Program: CLC 046 | Climate Adaptation for Program Managers (FAC 095) |
| Qualified Recycling Program Management: WENV 160 | MS in Systems Acquisition Management: 816 (degree) | Contracting Officers Representative (COR) Online Training: QC 222 | DoD Sustainable Procurement Program: CLC 046, also at DAU |
| Hazardous Materials Management Process (HMMP): WEW 222 | Acq Production, Quality and Manufacturing Decision Science: MN3384 | Life Cycle Logistics Foundational Level: LOG 105 | Partnership with more than 45 commercial providers, Project Management Institute (PMI), University of Virginia |
| Environmental Management in Deployed Locations: WENV 175 | Principles of Acq and Contract Management: MN3303 | Engineering and Technical Management Practitioner Level: ETM | Provides a list of tools for PMS and Contracting |
| Environmental Restoration Project Management: WENV 417 | Acq Management and Contract Admin: MN3315 | Test and Evaluation (T&E) Practitioner Level: TST 2040 | |
| Hazardous Waste Management: WEW 521 | Contracting for Major Systems: MN4304 | Partnership with Coursera and LinkedIn Learning | |
| Air Quality Management: WENV 531 | Contracting for Services: MN4311 | | |
| Water Quality Management: WEW 541 | | | |

Course with a focus on environmental consideration

General knowledge course, unit, or degree that includes environmental content

NOTE: Course names were compiled through course catalog reviews and subject-matter expert discussions. Acq = acquisition. Admin = administration. MBA = master of business administration. MS = master of science.

tion workforce and can be used in multiple ways, including by being applied toward a degree or certification or for ongoing training.[7]

---

[7] It is important to note that some in the DoD workforce are hired or transferred with preexisting expertise and bring that knowledge to their acquisition-related positions. Such expertise is important by itself but can also be passed on to colleagues over time.

## Resources and Tools for the Acquisition Workforce on Environmental Considerations

In addition to available formal education, training, and mentoring opportunities, other resources and tools exist to assist the acquisition workforce in incorporating environmental considerations into acquisition decisions. Some of these resources are offered by DoD, but many others are offered by other government agencies, such as the EPA, GSA, and DoE. NGOs (e.g., the Global Electronics Council) also provide public resources that can be accessed by DoD. Moreover, because environmental matters are always evolving, many of the available online resources, programs, and tools are continually refreshed to reflect current guidance.

One particularly notable resource available to DoD's personnel is the DENIX platform, which is a central site for installations, energy, environment, safety, and occupational health information. This online resource has drop-down menus for the categories of environment, safety, chemicals, energy, training, and archival materials.[8] Through this portal, personnel can access such resources as information on chemical and materials risk management, the National Defense Center for Energy and Environment, and ESOH in acquisition. While many of DENIX's resources are focused on traditional ESOH compliance, some are directed specifically to acquisition professionals. For example, one resource, *Sustainability Analysis Guidance: Integrating Sustainability into Acquisition Using Life Cycle Assessment*, " . . . presents a standardized framework for conducting a Sustainability Analysis, an assessment of costs (quantified using life cycle costing, LCC) and potential environmental liabilities (quantified using life cycle assessment, LCA) for DoD weapons systems, equipment, or platforms."[9]

One of the most comprehensive non-DoD resources is the Federal Facilities Environmental Stewardship and Compliance Assistance Center (FedCenter), which is a repository of environmentally oriented resources for federal purchasers and building managers, manufacturers, and the public.

[8] DENIX, homepage, undated-a.

[9] DoD, *Sustainability Analysis Guidance: Integrating Sustainability into Acquisition Using Life Cycle Assessment*, June 24, 2020, p. 5.

FedCenter is a joint initiative managed by the EPA Office of Enforcement and Compliance Assurance, the Army Corps of Engineers' Construction Engineering Research Laboratory, and the Office of the Federal Environmental Executive.[10] In fact, many of the resources and tools analyzed for this report were accessed from FedCenter's Acquisition section.

Tables 5.1 is a nonexhaustive list of resources available to the federal acquisition workforce. Many of the resources listed in Table 5.1 offer policy information hubs, catalogs of products that comply with a given program (e.g., the U.S. Department of Agriculture's BioPreferred Program), and sample contract language.

**TABLE 5.1**

**Sample of Environmentally Oriented Resources Available to Acquisition Professionals**

| Name | Description | Audience | Agency |
|---|---|---|---|
| FedCenter: Acquisition | Compilation of environmentally oriented acquisition material spanning government agencies and major NGOs | Federal acquisition workforce | EPA, USACE, Office of Federal CSO |
| DoD Sustainable Products Center | Independent virtual center integrating information on sustainable technologies | DoD ESOH, acquisition, and research personnel | DoD |
| DENIX | Virtual platform with links to such resources as UL SPOT, a database of environmentally certified products | ESOH personnel, regional environmental coordinators | DoD |
| Sustainable Facilities Tool | Comprehensive tool with information and resources to help meet high-performance planning, designing, and procurement needs | Federal facility managers, procurement professionals, leasing specialists, and project managers | GSA |
| Green Procurement Compilation | Comprehensive green purchasing resource with information about products, regulations, and a DoD-specific page | Federal acquisition professionals and program managers | GSA |

[10] FedCenter, "About FedCenter," webpage, undated-a.

**Table 5.1—Continued**

| Name | Description | Audience | Agency |
| --- | --- | --- | --- |
| Sustainable Marketplace | EPA initiative to provide information for federal purchasers looking to purchase greener products and services | Federal purchasers, consumers, industry | EPA |
| GSA Advantage! Environmental Program Aisle | Federal procurement site with thousands of options for products that are compliant with major environmental standards | Federal purchasers | GSA |
| Energy Efficient Product Procurement Program | FEMP site with information about energy-efficient products; includes case studies, a product catalog, and contract language | Federal purchasers, manufacturers, and vendors | FEMP |
| BioPreferred Program | Promotes use of biobased products (e.g., aircraft cleaners) via mandatory federal purchasing requirements and voluntary labeling opportunities | Federal agencies and contractors, corporations, general public | USDA |

SOURCES: FedCenter, "Acquisition," webpage, undated-b; DENIX, undated-e; DENIX, undated-a; Sustainable Facilities Tool, homepage, undated-a; Sustainable Facilities Tool, "Green Procurement Compilation," webpage, undated-b; EPA, "Sustainable Marketplace," webpage, updated April 19, 2023; GSA Advantage! "GSA Environmental Program Aisle," webpage, undated; Federal Energy Management Program, "Energy-Efficient Product Procurement," webpage, undated-a; U.S. Department of Agriculture, "BioPreferred Program," webpage, undated.

NOTES: CSO = Chief Sustainability Officer. FEMP = Federal Energy Management Program. USACE = U.S. Army Corps of Engineers. USDA = U.S. Department of Agriculture.

Table 5.2 shows a partial list of tools that are tailored to narrower uses than the resources listed in Table 5.1. The tools tend to be of "input-output" style (i.e., users can input criteria or technical information and the tool will produce one or more outputs).

As an example, the Defense Climate Assessment Tool (DCAT) can help acquisition professionals plan for the effects of a changing climate. DCAT was developed by the U.S. Army Corps of Engineers to allow installation planners and others to assess potential exposure to natural hazards affected by climate change. Initially developed for the Army, DCAT was adopted by OSD in 2019 to help ensure that DoD and the armed services evaluate vulnerability to natural hazards consistently. The tool projects hazards for two future periods—2050 and 2085—and two climate scenarios. Projected hazards are coastal flooding, riverine flooding, drought, energy demand, heat, historical weather extremes (e.g., tornado, hurricane-level wind, precipitation, ice storm or jam), land degradation, and wildfire. Exposure to these eight hazards is assessed and scored based on historical and projected data (where applicable) for a weighted set of normalized indicators that measure contributors to the hazard.[11] According to DoD personnel, DCAT could be useful to acquisition personnel considering long-term services or privatization contracts at military installations and could help enable installation master plans to satisfy a statutory mandate to address climate resilience.[12]

---

[11] A. O. Pinson, K. D. White, E. E. Ritchie, H. M. Conners, and J. R. Arnold, *DoD Installation Exposure to Climate Change at Home and Abroad*, U.S. Army Corps of Engineers, April 2021.

[12] According to the Deputy Assistant Secretary of Defense (Environment and Energy Resilience),

> The National Defense Authorization Act for Fiscal Year 2020 amended 10 USC Section 2864 to require that installation master plans address risks and threats to installation resilience, including those from climate change. The Department's September 2020 update to UFC 2-100-01, *Installation Master Planning* directs installations to incorporate climate resilience analysis in master planning activities to ensure mission sustainment over the intended lifespan of infrastructure and assets. The UFC also provides instruction on the use of climate scenario planning, and refers to the DoD Climate Assessment Tool (DCAT) and the DoD Regional Sea Level Database (DRSL). (Richard G. Kidd, IV, Deputy Assistant Secretary of Defense [Environment and Energy Resilience], statement before the Senate Committee on Appropriations, Subcommittee on Military Construction and Veterans Affairs, Military Infrastructure and Climate Resilience, May 19, 2021, p. 5)

**TABLE 5.2**

## Sample of Environmentally Oriented Tools Available to Acquisition Professionals

| Name | Description | Audience | Affiliated Organization |
|---|---|---|---|
| Acquisition Regulation Comparator | Online tool that enables acquisition personnel to compare up to three acquisition policies side by side (e.g., FAR versus DFARS) | Contracting officers, program managers, other acquisition personnel | GSA/DAU |
| EPEAT Benefits Calculator | Enables purchasers to calculate environmental benefits resulting from EPEAT products and estimate life-cycle impacts of various product-management strategies | Federal purchasers, private sector | Global Electronics Council |
| Federal Energy Management Tools | Features dozens of tools, including software, calculators, data sets, and databases designed to help federal agencies reduce energy use and achieve regulatory compliance | Federal agencies | FEMP |
| Life-Cycle Assessment Pave Tool | Voluntary tool that can aid life cycle accounting of environmental impacts of pavement and other design decisions | Federal agencies | DOT |
| DCAT | Calculates historical and projected installation exposure to eight natural hazards (e.g., coastal flooding, wildfires) to aid in incorporating environmental considerations into facilities and infrastructure planning and investments | Installation managers, DoD planners | DoD |

SOURCE: GSA and DAU, "Acquisition Regulation Comparator (ARC)," webpage, undated; Global Electronics Council, "Benefits Calculators," webpage, undated; Federal Energy Management Program, "Federal Energy Management Tools," webpage, undated-b; U.S. Department of Transportation, Federal Highway Administration, "LCA Pave Tool," webpage, updated April 4, 2022; DoD, *DoD Climate Assessment Tool*, fact sheet, undated.

NOTE: EPEAT = FEMP = Federal Energy Management Program. DOT = U.S. Department of Transportation.

CHAPTER 6

# Findings and Recommendations

To incorporate environmental considerations across the variety of both environmental issues and acquisition activities identified in Section 873 of the FY 2022 NDAA—covering environmental management, impact, compliance, resilience, and adaptation—the DoD acquisition workforce needs policies and guidance that tell them what to comply with, how to comply, and where and how in the requirements and acquisition process to engage. The acquisition workforce needs resources (e.g., SMEs and websites with information, useful links, models, and other tools) that provide information on the environmental and operational performance of systems, subsystems, components, and goods and services. The acquisition workforce also needs an understanding of technology that is currently in development that might mitigate environmental impacts (e.g., carbon emissions) or the impacts of the environment (e.g., extreme weather) on systems and goods and services. Finally, the acquisition workforce needs training and education on a variety of environmental considerations as part of their functional acquisition training.

In fact, these conditions currently exist in the acquisition community; DoD has been incorporating environmental considerations into acquisition planning and decisionmaking for at least several decades.[1]

A previous RAND study of environmental management conducted for DoD identified the following best practices applicable to DoD, including design-for-environment:

---

[1] As noted earlier, this research did not assess the sufficiency of DoD environment-related activities relative to its need or demand. In particular, we did not assess the resource sufficiency (staffing levels, funding) of the various organizations within DoD that are responsible for addressing environmental compliance and performance.

- Successful environmental management occurs when environmental concerns are linked to the mission objectives and integrated within organizational functions.
- Outcomes or successes rely on the individual decisionmaker, who must have the awareness, knowledge, and tools to consider environmental issues in decisions.
- Decisions should be integrated into existing processes seamlessly, with minimal burden and clear rationale.
- Individual decisionmakers need training and tools to increase awareness and knowledge and to facilitate analysis and decisionmaking.
- Organizational structures must be in place that address "environmental issues strategically and proactively, establish goals and communicate progress toward these goals, and create and share new knowledge about methodologies."[2]

DoD's acquisition-related environmental activities and initiatives largely reflect these best practices for incorporating environmental considerations into acquisition planning, analysis, and decisionmaking.

In defense acquisition, environmental performance and compliance are framed in terms of contribution to the operational mission. In fact, DoD policy recognizes environmental performance and compliance as improving readiness, resilience, and operational performance, thus emphasizing that compliance is not just for compliance's sake—it has a positive effect on mission performance and personnel. Furthermore, there is recognition in the department that improved environmental performance might result from other performance attributes (e.g., vehicle electrification reduces the demand for fuel and decreases noise and heat signature). Ultimately, environmental performance and compliance are treated like any other requirement within DoD's acquisition processes, which must necessarily prioritize military mission performance.

At a more detailed level, environmental considerations are integrated into existing decision points in the various acquisition processes. The policy and guidance in place emphasize the life cycle of a good or service from raw materials to disposal, ESOH, and climate resiliency. DoD culture is

---

2 Resetar et al., 1998, p. xvi.

already attuned to considering life cycle analysis (e.g., life cycle costs, sustainment, supply chain risk) in acquisition decisions. DoD policy has added sustainability to that list of life-cycle considerations in the acquisition of goods (commodities/supplies) and services,[3] while ESOH considerations are incorporated into the acquisition process for weapon systems.[4] Separate policy mandates that adaptation and resilience to climate change also be considered in mission planning and execution.[5]

DoD leverages and adopts commercial and industrial environmental standards and certifications, which is appropriate for the acquisition of goods and services. Environmental organizations at both the OSD and service levels are engaged with industry groups, other federal agencies (e.g., EPA, DoE, GSA), and the White House (e.g., Council on Environmental Quality), which allows DoD to maintain awareness of resources (e.g., information, certification programs, contract vehicles) and the latest in environmental technology and standards.

---

[3] DoDI 4105.72, 2018.

[4] DoDD 4715.1E, *Environment, Safety, and Occupational Health (ESOH)*, U.S. Department of Defense, December 30, 2019; OUSD(R&E), 2022b.

[5] DoDD 4715.21, *Climate Change Adaptation and Resilience*, U.S. Department of Defense, January 14, 2016. Per section 2.4 of Department of Defense Directive 4715.21 (pp. 5–6),

> Under the authority, direction, and control of the USD(AT&L), the Assistant Secretary of Defense for Acquisition:
>
> a. In accordance with DoDD 5000.01 and DoDI 5000.02:
>
> (1) Oversees integration of climate change considerations, including life cycle analyses, into acquiring or modifying weapons systems, platforms, equipment, and products.
>
> (2) Develops or updates policies to integrate climate change considerations into mission area analyses and acquisition strategies across the life cycle of weapons systems, platforms, and equipment.
>
> (3) Oversees integration of climate change-related policy and practices in defense acquisition workforce training and education.
>
> b. Supports the [Chairman of the Joint Chiefs of Staff] in integrating climate change considerations into interactions between the Joint Capabilities Integration and Development System and Defense Acquisition System processes in accordance with CJCS Instruction 3170.01I.

There is explicit recognition within the DoD acquisition community that "design in" provides greater flexibility and enhanced performance in meeting environmental standards and achieving environmental goals. This means incorporating environmental preferences in the requirements process and applies to both weapon system acquisition and the acquisition of goods and services.

Training and knowledge resources are available throughout DoD, the federal government, and the commercial sector. Environmental training and education is offered at DAU, AFIT, NPS, and other educational institutions within DoD, including both focused environmental-related courses and courses that include environmental considerations as part of other topics (e.g., civil engineering, product support). Training and education are also available to DoD personnel at both public and private schools, as well as at other supporting organizations (e.g., Federal Acquisition Institute).

## Key Findings

Our key findings are as follows:

1. **The DoD acquisition workforce appears to have the knowledge and tools to incorporate environmental considerations into acquisition planning, practice, and decisionmaking for weapon systems, goods, and services.** Policies and guidebooks are largely in place, as discussed in Chapters 2–4. Environmental organizations exist with the subject-matter expertise needed to support implementation of those policies. DoD websites contain resources and links to additional information and requirements that developers, program managers, and contracting officers need to guide implementation of those policies. Relevant training and education is available to the acquisition community at internal DoD educational institutions and at public and private external learning organizations. Moreover, DoD has been implementing environmental activities and initia-

tives that are relevant to acquisition for many decades, so the workforce is building on a foundation of experience.[6]

2. **DoD has long-standing policy and guidance in many areas related to environmental considerations; however, there is a potential gap in environment-specific functional policy and guidance.** These policies, which we discussed in Chapter 4, demonstrate awareness of the knowledge and tools needed to incorporate environmental considerations into acquisition planning and practice. There are policies specific to environmental considerations, and the requirement to address environmental issues is stated in many acquisition policies. Some functional policy and related guidance also address environmental considerations. However, there is no functional policy or guidance focused solely on environmental management, compliance, impact, and related issues. This is potentially a gap.
3. **For weapon systems, environmental considerations are incorporated in system engineering, design interface, and product support processes and treated like any other performance or compliance requirement.** While most acquisition policies (see Chapter 4) for weapon system acquisition—for example, the DoDI 5000 series and AAF pathways—mention the need for ESOH compliance, the "how to" steps are integrated into the broader systems engineering function and documented in some detail in the *Systems Engineering Guidebook*. Other acquisition-related guidebooks also include some guidance on how environmental considerations should be addressed as part of the acquisition or sustainment process, including the Product Support Manager and Test and Evaluation Guidebooks. In policy and guidance, environmental compliance is treated the same way as other functional areas requiring compliance (e.g., cybersecurity), and environmental performance becomes a requirement that is evaluated with and against other performance requirements in the context of mission performance, cost, and schedule.

[6] Again, we did not evaluate whether and are not arguing that DoD's environmental initiatives are sufficient or well-designed and implemented; rather, DoD has the knowledge and tools needed to incorporate environmental considerations into acquisition planning, practice, and decisionmaking, and has been doing so.

4. **Policy on sustainability and FAR Part 23 include rules for incorporating environmentally friendly preferences in procurement of goods and services and leverage commercial and industry standards.** As noted in Chapter 4, policy and regulations require consideration of environmental compliance, impact, and sustainability, and either provide or point to the knowledge and tools required to do so. Commodity procurement (goods/supplies) and acquisition of services are different processes than weapon system acquisition. The policy that sets procedures and roles and responsibilities for the procurement of sustainable goods and services—DoDI 4015.72—is different from the policies for weapon systems. FAR Part 23 also provides a regulatory framework for incorporating sustainability into acquisition of goods and services. For goods or services that are essentially commercial products, DoD can and does adopt and leverage commercial or industry standards as employed in the commercial marketplace. This includes leveraging environmental performance and energy efficiency product certifications issued by other federal agencies (e.g., EPA, DoE).
5. **Participation in internal DoD forums (e.g., the Climate Working Group) and interagency forums enables the DoD acquisition community to be aware of and leverage knowledge and tools within DoD and other federal agencies.** Within DoD, OSD and service organizations that are responsible for incorporating environmental considerations into acquisition planning and decisionmaking work together and share information. They also participate in interagency working groups at the federal level and with industry groups. This allows DoD to rely on commercial standards or government certification for environmental performance for goods and services.
6. **The DoD acquisition workforce has access to a variety of educational options to improve workforce knowledge of incorporating environmental considerations in acquisition.** To support the acquisition workforce, multiple institutions develop and provide educational opportunities and training on environmental considerations. DoD and federal education institutions, private colleges and universities, and commercial-sector partnerships all offer courses that are accessible to DoD's workforce. Most courses focus on environmental

considerations in the context of a broader topic, such as program management or contracting, but there are courses that are focused on environmental considerations for specific degrees or focus areas, such as civil engineering. Along with courses through outside institutions and those in partnership with DoD, specific offices focused on environmental considerations within DoD also provide training. In addition to formalized courses and training opportunities available through these and other institutions, many more resources and tools exist to assist the acquisition workforce in incorporating environmental considerations into their job duties. Because environmental matters are continually evolving, many online resources, programs, and tools are available for the workforce to consult to supplement continued learning about environmental topics. Additional resources might be available in the future; the CLSWG is currently assessing climate literacy across the DoD workforce and will be recommending "means and methods for tailoring and/or improving climate education, training, and engagement."[7]

In summary, the knowledge and tools required to incorporate environmental cost, resource, and energy efficiency and resilience considerations exist in DoD's environmental organizations, and policy and regulations direct how and when those environmental considerations should be input into acquisition planning and process generally and into source selection specifically. DoD also has several R&D programs that invest in and demonstrate environmentally friendly technologies.

## Recommendations

Our recommendations to improve the incorporation of environmental considerations into acquisition planning and decisionmaking **build on existing DoD environmental capacity, capabilities, and activities.**

[7] Assistant Secretary of Defense for Readiness, "Climate Literacy Sub-Working Group Overview," undated, p. 6. This briefing was provided to the authors by the Deputy Assistant Secretary of Defense (Force Education and Training) and is not available to the general public.

There are many environmental initiatives and compliance activities at all levels of DoD, mostly in the background. Existing organizations have the subject-matter expertise needed to design policies and advise leadership on the full variety of environmental issues. Organizations within the services' life-cycle management or system commands act as environmental advisers to weapon system programs, particularly for smaller programs. Building on this structure, DoD should formalize the role of environmental adviser to acquisition activities. **Creating an environmental adviser function** (similar to a small business adviser) to assist in the procurement of environmentally preferred goods and services would increase the visibility of DoD environmental policy and help achieve DoD's environmental goals. The environmental adviser, whether for weapon system acquisition or procurement of goods and services, should engage early in the requirements process where "design in" decisions can have a greater impact.

We identified several areas in which environmental policy and guidance could be strengthened, particularly in areas related to weapon system acquisition. Along with maintaining and updating the environmental requirements in acquisition policy and the "how to" in existing functional guidance, we suggest **creating and maintaining an environmental guidebook** as a resource for requirements developers, program managers, contracting officers, and others in the acquisition community. The intent is not to replace the guidance in existing functional policy and guidebooks, but rather to bring together the knowledge and lessons of environmental management as applied in acquisition processes. Functional guidance is also easier to update than policy in response to changes in technology or environmental impacts that require changes in environmental practice.

We also recommend **continuing and enhancing collaboration and information-sharing across DoD** and with other federal, state, local, and industry organizations. This could include establishing a central repository of environmental performance data and other information, including lessons from past and ongoing initiatives and technology demonstrations.

Finally, building on existing practice in GSA and other federal contracting organizations, DoD should **establish task order general contract vehi-**

**cles with prequalified firms for select environmentally preferred goods and services** (in addition to what is already provided by GSA).[8]

## DoD Environmental Initiatives Would Benefit from Further Research

We also identified areas in which further research would benefit DoD environmental initiatives:

- **Conduct resource analysis to identify staffing and funding needs.** This work should focus on enhancing DoD's capacity and capability to incorporate environmental considerations into acquisition planning and processes. This should include funding availability for R&D and technology demonstration and testing, which was an issue during multiple discussions with SMEs.
- **Examine implementation of current initiatives to better understand enablers and mitigate challenges and barriers.** Although DoD has been incorporating environmental considerations into acquisition planning and decisionmaking for several decades, close examination of how those initiatives have been implemented and the challenges they face would inform and enhance future DoD environmental policy design and implementation.
- **Explore the creation of an environmental specialist career category,** either as a stand-alone career or within one or more existing fields (e.g., engineering, logistics) in the acquisition workforce. This could be one way to raise the visibility of environmental issues within the DoD workforce and could facilitate the recruitment of environmental scientists and related talent.
- **Explore the utility of creating a cross-cutting acquisition policy on environment** (similar to DoDI 5000.90 on cybersecurity). Environmental and energy-related issues touch everything that DoD does. Environmental considerations cut across many other policy areas and activities, with implications for readiness, resilience, efficiency, and

[8] DoD recently did something similar. See DoD, 2023.

effectiveness. A cross-cutting environmental policy raises the visibility of the issue across the DoD enterprise and can contribute to changes in behavior and culture that might accelerate DoD's ability to adapt to changes in the operational environment and mitigate environmental impacts.

- **Explore the utility of establishing one or more environmental KPPs to be included in requirements processes.** An energy KPP already exists and results in energy efficiency to be considered in system performance. Although environmental performance might be difficult to measure (i.e., How can the impact of reduced use of hazardous chemicals on readiness be measured?), it is possible that some useful environmental performance standards can be established that help inform performance trade-offs usually done in the requirements and design phases.
- **Explore establishing environmental data governance and management processes**, develop additional metrics, and build on existing analytic capabilities to inform acquisition planning and decisionmaking and track impact on readiness, resilience, and operations. DoD's current acquisition data governance practices represent a model that is applicable to environmental information.

Many of these ideas for further research came up as discussion topics in interviews or are the result of observations made by the research team. The resource analysis and implementation constraints ideas came up in multiple interviews and reflect the perceptions of officials working in the environmental space. Creating an environmental specialist career track might help DoD attract the talent needed to address environmental compliance and impact. A cross-cutting functional environmental policy would complement the environmental guidebook we recommend. Finally, environmental KPPs and formalizing data governance associated with environmental reporting can help DoD understand its environmental impact and inform future decisions regarding incorporating environmental considerations into acquisition practice and decisionmaking.

# Abbreviations

| | |
|---|---|
| AAF | Adaptive Acquisition Framework |
| AFIT | Air Force Institute of Technology |
| ASA(IE&E) | Assistant Secretary of the Army (Installations, Energy and Environment) |
| ASD(A)/AE | Assistant Secretary of Defense (Acquisition)/Acquisition Enablers |
| ASN(EI&E) | Assistant Secretary of the Navy (Energy, Installations and Environment) |
| CLSWG | Climate Literacy Sub-Working Group |
| DAU | Defense Acquisition University |
| DCAT | Defense Climate Assessment Tool |
| DENIX | U.S. Department of Defense Environment, Safety and Occupational Health Network and Information Exchange |
| DFARS | Defense Federal Acquisition Regulation Supplement |
| DIU | Defense Innovation Unit |
| DoD | U.S. Department of Defense |
| DoDD | Department of Defense Directive |
| DoDI | Department of Defense Instruction |
| DoE | U.S. Department of Energy |
| EO | executive order |
| EPA | U.S. Environmental Protection Agency |
| ESOH | Environmental, Safety, and Occupational Health |
| ESTCP | Environmental Security Technology Certification Program |
| FAR | Federal Acquisition Regulation |
| FedCenter | Federal Facilities Environmental Stewardship and Compliance Assistance Center |
| FFRDC | federally funded research and development center |
| FY | fiscal year |
| GSA | General Services Administration |

| | |
|---|---|
| HAZMAT | hazardous materials |
| KPP | Key Performance Parameter |
| LOE | lines of effort |
| NDAA | National Defense Authorization Act |
| NDCEE | National Defense Center for Energy and Environment |
| NDS | National Defense Strategy |
| NEPA | National Environmental Policy Act |
| NGO | nongovernmental organization |
| NPS | Naval Postgraduate School |
| ODASD | Office of the Deputy Assistant Secretary of Defense |
| ODASD (E&ER) | Office of the Deputy Assistant Secretary of Defense (Environment and Energy Resilience) |
| ODASD (Energy) | Office of the Deputy Assistant Secretary of Defense (Energy Directorate) |
| OMB | U.S. Office of Management and Budget |
| OSD | Office of the Secretary of Defense |
| OUSD(A&S) | Office of the Under Secretary of Defense for Acquisition and Sustainment |
| OUSD(R&E) | Office of the Under Secretary of Defense for Research and Engineering |
| PESHE | Programmatic Environmental, Safety, and Health Evaluation |
| PFAS | per- and polyfluoroalkyl substances |
| R&D | research and development |
| RDT&E | research, development, test, and evaluation |
| SAF/IE | Office of the Assistant Secretary of the Air Force for Energy, Installations, and Environment |
| SAF/IEE | Office of the Assistant Secretary of the Air Force for Environment, Safety, and Infrastructure |
| SAF/IEI | Office of the Assistant Secretary of the Air Force for Installations |
| SAF/IEN | Office of the Assistant Secretary of the Air Force for Operational Energy |

| | |
|---|---|
| SERDP | Strategic Environmental Research and Development Program |
| SME | subject-matter expert |

# References


AcqNotes, "Program Management: Programmatic Environmental Safety and Occupational Health Evaluation," webpage, June 14, 2018. As of February 13, 2023:
https://acqnotes.com/acqnote/careerfields/programmatic-environmental-safety-and-occupational-health-evaluation

Air Force Institute of Technology, Graduate School of Engineering and Management, *Graduate School Academic Catalog 2021–2022*, undated.

Air Force Life Cycle Management Center, "AFLCMC Focus Week Course Catalog, 25–29 April 2022," briefing slides, March 29, 2022.

Air Force Operational Energy, *Annual Report 2020*, Office of the Assistant Secretary of the Air Force for Energy, Installations, and Environment, undated.

Army Directive 2020-03, *Installation Energy and Water Resilience Policy*, U.S. Department of Defense, March 31, 2020.

Army Directive 2020-11, *Roles and Responsibilities for Military Installation Operations*, U.S. Department of Defense, October 1, 2020.

ASA(IE&E)—*See* Assistant Secretary of the Army (Installations, Energy and Environment).

ASN(EI&E)—*See* Assistant Secretary of the Navy (Energy, Installations and Environment).

Assistant Secretary of the Army (Installations, Energy and Environment), "About Us" webpage, undated-a. As of February 15, 2023:
https://www.army.mil/asaiee#org-about-us

Assistant Secretary of the Army (Installations, Energy and Environment), "Helpful Links to ASA(IE&E) Directorates, Programs and the Installation Management Community," webpage, undated-b. As of February 15, 2023:
https://www.army.mil/asaiee#org-ie-e-info-links

Assistant Secretary of the Army (Installations, Energy and Environment), "Sustainability," webpage, updated September 2021. As of February 15, 2023:
https://www.asaie.army.mil/Public/ES/sustainability.html

Assistant Secretary of the Navy (Energy, Installations and Environment), "Deputy Assistant Secretary of the Navy (Environment and Mission Readiness)," webpage, undated-a. As of February 15, 2023:
https://www.secnav.navy.mil/eie/Pages/Environment.aspx

Assistant Secretary of the Navy (Energy, Installations and Environment), "Deputy Assistant Secretary of the Navy (Installations, Energy and Facilities)," webpage, undated-b. As of February 15, 2023:
https://www.secnav.navy.mil/eie/Pages/InstallationsFacilities.aspx

Assistant Secretary of the Navy (Energy, Installations and Environment), "Deputy Assistant Secretary of the Navy (Safety)," webpage, undated-c. As of February 15, 2023:
https://www.secnav.navy.mil/eie/Pages/Safety.aspx

Best, Katharina Ley, Scott R. Stephenson, Susan A. Resetar, Paul W. Mayberry, Emmi Yonekura, Rahim Ali, Joshua Klimas, Stephanie Stewart, Jessica Arana, Inez Khan, and Vanessa Wolf, *Climate and Readiness: Understanding Climate Vulnerability of U.S. Joint Force Readiness*, RAND Corporation, RR-A1551-1, 2023. As of May 24, 2023:
https://www.rand.org/pubs/research_reports/RRA1551-1.html

Biden, Joseph R., "Executive Order on Tackling the Climate Crisis at Home and Abroad," Executive Order 14008, Executive Office of the President, January 27, 2021a.

Biden, Joseph R., "Executive Order on Catalyzing Clean Energy Industries and Jobs Through Federal Sustainability," Executive Order 14057, Executive Office of the President, December 8, 2021b.

Camm, Frank, Jeffrey A. Drezner, Beth E. Lachman, and Susan A. Resetar, *Implementing Proactive Environmental Management: Lessons Learned from Best Commercial Practice*, RAND Corporation, MR-1371-OSD, 2001. As of January 31, 2023:
https://www.rand.org/pubs/monograph_reports/MR1371.html

Chairman of the Joint Chiefs of Staff Instruction 5123.011, *Charter of the Joint Requirements Oversight Council and Implementation of the Joint Capabilities Integration and Development System*, October 30, 2021.

Conger, John, "And Air Force Makes Three . . . Comparing the U.S. Army, Navy and Air Force Climate Plans," Center for Climate and Security, October 5, 2022.

Cook, Cynthia R., Éder Sousa, Yool Kim, Megan McKernan, Yuliya Shokh, Sydne J. Newberry, Kelly Elizabeth Eusebi, and Lindsay Rand, *Ensuring Mission Assurance While Conducting Rapid Space Acquisition*, RAND Corporation, RR-A998-1, 2022. As of January 31, 2023:
https://www.rand.org/pubs/research_reports/RRA998-1.html

DAU—*See* Defense Acquisition University.

Defense Acquisition University, "Acquisition Functional Areas," webpage, undated-a. As of December 3, 2022:
https://www.dau.edu/functional-areas

Defense Acquisition University, "Adaptive Acquisition Framework Pathways: Acquisition Policies," webpage, undated-b. As of February 15, 2023: https://aaf.dau.edu/aaf/policies/

Defense Acquisition University, "Adaptive Acquisition Framework Pathways: Acquisition of Services," webpage, undated-c. As of February 9, 2023: https://aaf.dau.edu/aaf/services/

Defense Acquisition University, "iCatalog Home Page," webpage, undated-d. As of February 13, 2023: https://icatalog.dau.edu/icatalog_home.aspx

Defense Acquisition University, "Systems Engineering Brainbook: ESOH Risk Assessment," webpage, undated-e. As of February 13, 2023: https://www.dau.edu/tools/se-brainbook/Pages/Management%20Processes/esoh-risk-assessment.aspx

Defense Acquisition University Service Acquisition Mall, "Service Acquisition Steps," webpage, undated. As of February 15, 2023: https://www.dau.edu/tools/Documents/SAM/home.html

DENIX—*See* U.S. Department of Defense Environment, Safety and Occupational Health Network and Information Exchange.

Department of the Air Force, Office of the Assistant Secretary for Energy, Installations, and Environment, *Climate Action Plan*, October 2022.

Department of the Army, Office of the Assistant Secretary of the Army for Installations, Energy and Environment, *United States Army Climate Strategy*, February 2022.

Department of Defense Directive 4715.1E, *Environment, Safety, and Occupational Health (ESOH)*, U.S. Department of Defense, December 30, 2019.

Department of Defense Directive 4715.21, *Climate Change Adaptation and Resilience*, U.S. Department of Defense, January 14, 2016.

Department of Defense Financial Management Regulation 7000.14-R, Volume 2B, Budget Formulation and Presentation, Chapter 5, September 2022.

Department of Defense Instruction 4105.72, *Procurement of Sustainable Goods and Services*, U.S. Department of Defense, September 7, 2016, change 1, August 31, 2018.

Department of Defense Instruction 5000.02, *Operation of the Adaptive Acquisition Framework*, U.S. Department of Defense, January 23, 2020, change 1, June 8, 2022.

Department of Defense Instruction 5000.88, *Engineering of Defense Systems*, U.S. Department of Defense, November 18, 2020.

Department of the Navy, Office of the Assistant Secretary of the Navy for Energy, Installations, and Environment, *Climate Action 2030*, May 2022.

Deputy Assistant Secretary of Defense for Force Education and Training, “About Us,” webpage, undated-a. As of April 10, 2023: https://prhome.defense.gov/Readiness/Organization/FET/

Deputy Assistant Secretary of Defense for Force Education and Training, “Programs,” webpage, undated-b. As of April 10, 2023: https://prhome.defense.gov/Readiness/Organization/FET/SDEF/Programs/

DoD—*See* U.S. Department of Defense.

DoDD—*See* Department of Defense Directive.

DoDI—*See* Department of Defense Instruction.

Drezner, Jeffrey A., and Melissa A. Bradley, *A Survey of DoD Facility Energy Management Capabilities*, RAND Corporation, MR-875-OSD, 1998. As of February 23, 2023: https://www.rand.org/pubs/monograph_reports/MR875.html

EPA—*See* U.S. Environmental Protection Agency.

Evans, Gareth, “US Green Fleet: A New Era of Naval Energy,” *Naval Technology*, May 3, 2016.

FedCenter—*See* Federal Facilities Environmental Stewardship and Compliance Assistance Center.

Federal Acquisition Institute, homepage, undated. As of February 13, 2023: https://www.fai.gov

Federal Acquisition Regulation, Part 23, Environment, Energy and Water Efficiency, Renewable Energy Technologies, Occupational Safety, and Drug-Free Workplace, March 16, 2023a.

Federal Acquisition Regulation, Part 23, Environment, Energy and Water Efficiency, Renewable Energy Technologies, Occupational Safety, and Drug-Free Workplace, Subpart 23.103, Sustainable Acquisitions, March 16, 2023b.

Federal Acquisition Regulation, Part 36, Construction and Architect-Engineer Contracts, Subpart 36.104, Policy, February 13, 2023c.

Federal Energy Management Program, “Energy-Efficient Product Procurement,” webpage, undated-a. As of February 15, 2023: https://www.energy.gov/femp/energy-efficient-product-procurement

Federal Energy Management Program, “Federal Energy Management Tools,” webpage, undated-b. As of February 15, 2023: https://www.energy.gov/eere/femp/federal-energy-management-tools

Federal Facilities Environmental Stewardship and Compliance Assistance Center, “About FedCenter,” webpage, undated-a. As of February 10, 2023: https://www.fedcenter.gov/help/about/

Federal Facilities Environmental Stewardship and Compliance Assistance Center, "Acquisition," webpage, undated-b. As of February 15, 2023: https://www.fedcenter.gov/programs/buygreen/

Frost, Christopher, "The Great Green Fleet Operates in the South China Sea," *PACOM News*, March 4, 2016.

Garamone, Jim, "DoD Office Focuses on Effects of Climate Change on Department," *Anchorage Press*, August 3, 2022.

General Services Administration and the Defense Acquisition University, "Acquisition Regulation Comparator (ARC)," webpage, undated. As of February 15, 2023:
https://www.acquisition.gov/arc

Global Electronics Council, "Benefits Calculators," webpage, undated. As of February 15, 2023:
https://globalelectronicscouncil.org/benefits-calculators/

GSA Advantage! "GSA Environmental Program Aisle," webpage, undated. As of February 15, 2023:
https://www.gsaadvantage.gov/advantage/ws/search/special_category_search?cat=ADV.ENV

Heise, Rene, "NATO Is Responding to New Challenges Posed by Climate Change," *NATO Review*, April 1, 2021.

Hicks, Kathleen H., Deputy Secretary of Defense, "Energy Supportability and Demand Reduction in Capability Development," memorandum for Secretaries of the Military Departments; Chairman of the Joint Chiefs of Staff; and Under Secretary of Defense for Acquisition and Sustainment, April 21, 2022.

Kidd, Richard G., IV, Deputy Assistant Secretary of Defense (Environment and Energy Resilience), statement before the Senate Committee on Appropriations, Subcommittee on Military Construction and Veterans Affairs, Military Infrastructure and Climate Resilience, May 19, 2021.

Koehler, Erv, "*Ready, Set, STED: Speeding Up Sustainable Acquisition*," *GSA Blog*, December 15, 2022. As of February 13, 2023:
https://www.gsa.gov/blog/2022/12/15/ready-set-sted-speeding-up-sustainable-acquisition

Kuzmitski, Holly, "Remote Sensing Gives USACE an Edge at Detecting Harmful Algal Blooms," webpage, U.S. Army Corps of Engineers, January 23, 2023. As of February 15, 2023:
https://www.usace.army.mil/Media/News/NewsSearch/Article/3277031/remote-sensing-gives-usace-an-edge-at-detecting-harmful-algal-blooms/

Mayfield, Mandy, "SOFIC NEWS: Pentagon Looks to Incorporate 'Climate Resilience' into Future Weapon Systems," *National Defense*, May 19, 2021.

McKernan, Megan, Jeffrey A. Drezner, Mark V. Arena, Jonathan P. Wong, Yuliya Shokh, Austin Lewis, Nancy Young Moore, Judith D. Mele, and Sydne J. Newberry, *Using Metrics to Understand the Performance of the Adaptive Acquisition Framework*, RAND Corporation, RR-A1349-1, 2022. As of January 31, 2023:
https://www.rand.org/pubs/research_reports/RRA1349-1.html

McNamara, Kelly, "ENV 101—Introduction to ESOH in Acquisition," briefing slides, U.S. Air Force Life Cycle Management Center, April 25, 2022.

Military Standard 882E, *Department of Defense Standard Practice: System Safety*, U.S. Department of Defense, May 11, 2012.

Naval Postgraduate School, "Department of Defense Management," webpage, undated. As of February 13, 2023:
https://nps.edu/web/ddm/acquisition-management

Navy Environmental Sustainability Development to Integration, *Assessment of Cadmium Alternatives for Connector Applications*, undated.

National Defense Center for Energy and Environment, *How to Do Business with NDCEE: A Guide for Our Stakeholders*, 2023.

NDCEE—*See* National Defense Center for Energy and Environment.

Nugee, Richard, "A Growing Crisis: The Launch of the World Climate and Security Report," Expert Group of the International Military Council on Climate and Security, June 7, 2021.

Office of the Assistant Secretary of the Air Force for Energy, Installations, and Environment, "About Us," webpage, undated-a. As of February 15, 2023:
https://www.safie.hq.af.mil/About-Us/

Office of the Assistant Secretary of the Air Force for Energy, Installations, and Environment, "Air Force Energy Program," webpage, undated-b. As of February 17, 2023:
https://www.safie.hq.af.mil/Energy

Office of the Assistant Secretary of the Air Force for Energy, Installations, and Environment, "Air Force Operational Energy About Us," webpage, undated-c. As of February 15, 2023:
https://www.safie.hq.af.mil/OpEnergy/About/

Office of the Assistant Secretary of the Air Force for Energy, Installations, and Environment, "Installations," webpage, undated-d. As of February 15, 2023:
https://www.safie.hq.af.mil/Programs/Installations/

Office of the Assistant Secretary of the Air Force for Energy, Installations, and Environment, "Office of Energy Assurance: Your Storefront for Creative Energy Solutions," website, undated-e. As of February 17, 2023:
https://www.safie.hq.af.mil/Programs/Energy/OEA

Office of the Assistant Secretary of the Air Force for Energy, Installations, and Environment, "SAF/IEE Environment, Safety, and Infrastructure: What We Do," webpage, undated-f. As of February 15, 2023:
https://www.safie.hq.af.mil/About-Us/Units/Environment-Safety-Infrastructure/

Office of the Assistant Secretary of the Army for Installations, Energy and Environment, "About Us" webpage, undated-a. As of February 15, 2023:
https://www.army.mil/asaiee#org-about-us

Office of the Assistant Secretary of the Army for Installations, Energy and Environment, "ESOH," webpage, undated-b. As of February 15, 2023:
https://www.asaie.army.mil/Public/ESOH/

Office of the Assistant Secretary of the Army for Installations, Energy and Environment, "Strategic Integration," webpage, undated-c. As of February 15, 2023:
https://www.asaie.army.mil/Public/SI/

Office of the Assistant Secretary of the Army for Installations, Energy and Environment, *Safety, Occupational and Environmental Health (SO&EH) Strategy 2020–2028*, U.S. Army, April 28, 2020.

Office of the Assistant Secretary of the Army for Installations, Energy and Environment, *ASA (IE&E): Installations, Energy and Environment—Fiscal Year 2021: Year in Review*, U.S. Army, October 1, 2021.

Office of the Assistant Secretary of Defense for Sustainment, "Deputy Assistant Secretary of Defense for Environment & Energy Resilience," webpage, undated-a. As of February 15, 2023:
https://www.acq.osd.mil/log/EER/index.html

Office of the Assistant Secretary of Defense for Sustainment, "Installation Energy," webpage, undated-b. As of February 13, 2023:
https://www.acq.osd.mil/eie/IE/FEP_index.html

Office of the Assistant Secretary of Defense for Sustainment, "Library, Resources & Archives," webpage, undated-c. As of February 15, 2023:
https://www.acq.osd.mil/eie/Library.html

Office of the Assistant Secretary of Defense for Sustainment, "Welcome to Energy," webpage, undated-d. As of February 15, 2023:
https://www.acq.osd.mil/log/ENR/index.html

Office of the Assistant Secretary of Defense for Sustainment, "Welcome to Environment," webpage, undated-e. As of February 15, 2023:
https://www.acq.osd.mil/log/ENV/index.html

Office of the Assistant Secretary of the Navy (Energy, Installations and Environment), *Fiscal Year 2020 Annual Report*, Department of the Navy, undated.

Office of the Deputy Assistant Secretary of Defense for Product Support, *Product Support Manager Guidebook*, U.S. Department of Defense, May 2011, updated May 24, 2022.

Office of the Deputy Under Secretary of Defense for Installations and Environment and Office of the Deputy Under Secretary of Defense for Acquisition and Technology, *Environment, Safety, and Occupational Health (ESOH) in Acquisition: Integrating ESOH into Systems Engineering*, April 2009.

Office of the Federal Chief Sustainability Officer, Council on Environmental Quality, "Department of Defense Agency Progress," webpage, undated-a. As of February 13, 2023:
https://www.sustainability.gov/dod.html

Office of the Federal Chief Sustainability Officer, Council on Environmental Quality, "Federal Progress, Plans, and Performance," webpage, undated-b. As of February 15, 2023:
https://www.sustainability.gov/progress.html

Office of the President, "Executive Order 13514—Federal Leadership in Environmental, Energy, and Economic Performance," *Federal Register*, Vol. 74, No. 194, October 8, 2009.

Office of the Secretary of Defense, *Department of Defense Fiscal Year (FY) 2023 Budget Estimates: Defense-Wide Justification Book Volume 3 of 5—Research, Development, Test & Evaluation, Defense-Wide*, April 2022.

Office of the Under Secretary of Defense for Acquisition and Sustainment, "Brendan Owens: Assistant Secretary of Defense for Energy, Installations, and Environment," webpage, undated-a. As of March 21, 2023:
https://www.acq.osd.mil/leadership/eie/brendan-owens.html

Office of the Under Secretary of Defense for Acquisition and Sustainment, "OUSD A&S Organizations," webpage, undated-b. As of April 10, 2023:
https://www.acq.osd.mil/organizations.html

Office of the Under Secretary of Defense for Acquisition and Sustainment, *Report on Effects of a Changing Climate to the Department of Defense*, U.S. Department of Defense, January 10, 2019.

Office of the Under Secretary of Defense for Acquisition and Sustainment, *Defense Environmental Programs Annual Report to Congress for Fiscal Year 2020*, March 2022a.

Office of the Under Secretary of Defense for Acquisition and Sustainment, *Department of Defense Sustainability Plan: 2022*, U.S. Department of Defense, March 2022b.

Office of the Under Secretary of Defense for Acquisition and Sustainment, *Department of Defense Climate Adaptation Plan: 2022 Progress Report*, U.S. Department of Defense, October 4, 2022c.

Office of the Under Secretary of Defense for Research and Engineering, Office of the Deputy Director for Engineering, *Engineering of Defense Systems Guidebook*, U.S. Department of Defense, February 2022a.

Office of the Under Secretary of Defense for Research and Engineering, Office of the Deputy Director for Engineering, *Systems Engineering Guidebook*, U.S. Department of Defense, February 2022b.

Office of the Under Secretary of Defense for Research and Engineering, Office of the Deputy Director for Engineering, *Human Systems Integration Guidebook*, U.S. Department of Defense, May 2022c.

Office of the Under Secretary of Defense for Research and Engineering and Office of the Director for Operational Test and Evaluation, *Test and Evaluation Enterprise Guidebook*, U.S. Department of Defense, August 2022.

OSD—*See* Office of the Secretary of Defense.

OUSD(A&S)—*See* Office of the Under Secretary of Defense for Acquisition and Sustainment.

OUSD(R&E)—*See* Office of the Under Secretary of Defense for Research and Engineering.

Pinson, A. O., K. D. White, S. A. Moore, S. D. Samuelson, B. A. Thames, P. S. O'Brien, C. A. Hiemstra, P. M. Loechl, and E. E. Ritchie, *Army Climate Resilience Handbook*, U.S. Army Corps of Engineers, August 2020.

Pinson, A. O., K. D. White, E. E. Ritchie, H. M. Conners, and J. R. Arnold, *DoD Installation Exposure to Climate Change at Home and Abroad*, U.S. Army Corps of Engineers, April 2021.

Public Law 117–81, National Defense Authorization Act for Fiscal Year 2022, December 27, 2021.

Rendon, Rene G., *Commodity Sourcing Strategies: Supply Management in Action*, Naval Postgraduate School, January 31, 2005.

Resetar, Susan A., Frank Camm, and Jeffrey A. Drezner, *Environmental Management in Design: Lessons from Volvo and Hewlett-Packard for the Department of Defense*, RAND Corporation, MR-1009-OSD, 1998. As of January 31, 2023:
https://www.rand.org/pubs/monograph_reports/MR1009.html

Saballa, Joe, "US Army Seeking All-Electric Vehicle Fleet to Slash Carbon Emissions," *Defense Post*, February 10, 2022.

SAF/IE—*See* Office of the Assistant Secretary of the Air Force for Energy, Installations, and Environment.

Samaras, Constantine, Abigail Haddad, Clifford A. Grammich, and Katharine Watkins Webb, *Obtaining Life-Cycle Cost-Effective Facilities in the Department of Defense*, RAND Corporation, RR-169-OSD, 2013. As of February 15, 2023: https://www.rand.org/pubs/research_reports/RR169.html

Sargent, John F., Jr., *Department of Defense Research, Development, Test, and Evaluation (RDT&E): Appropriations Structure*, Congressional Research Service, R44711, September 7, 2022.

Secretary of Defense, "Establishment of the Climate Working Group," memorandum for Senior Pentagon Leadership, Commanders of the Combatant Commands, and Defense Agency and DoD Field Activity Directors, U.S. Department of Defense, March 9, 2021.

SERDP and ESTCP—*See* Strategic Environmental Research and Development Program and Environmental Security Technology Certification Program.

Strategic Environmental Research and Development Program and Environmental Security Technology Certification Program, homepage, undated-a. As of February 13, 2023:
https://www.serdp-estcp.org/

Strategic Environmental Research and Development Program and Environmental Security Technology Certification Program, "About Us," webpage, undated-b. As of February 17, 2023:
https://www.serdp-estcp.org/about

Sustainable Facilities Tool, homepage, undated-a. As of February 15, 2023: https://sftool.gov

Sustainable Facilities Tool, "Green Procurement Compilation," webpage, undated-b. As of February 15, 2023:
https://sftool.gov/greenprocurement

Tiedeman, Melissa, "Year in Review: 2018 SAF/IEE Installation Energy," Office of the Assistant Secretary of the Air Force for Energy, Installations, and Environment, December 20, 2018.

U.S. Army, *Army Installation Energy and Water Strategic Plan*, December 2020.

U.S. Code, Title 10, Section 2901, Strategic Environmental Research and Development Program, to Section 2904, Strategic Environmental Research and Development Program Scientific Advisory Board.

U.S. Code, Title 10, Section 2911, Energy Policy of the Department of Defense.

U.S. Department of Agriculture, "BioPreferred Program," webpage, undated. As of February 15, 2023:
https://www.biopreferred.gov/BioPreferred/

U.S. Department of the Air Force organizational chart, AFVA 38-104, supersedes AFVA 38-104, November 16, 2022. As of April 12, 2023: https://static.e-publishing.af.mil/production/1/saf_aa/publication/afva38-104/afva38-104.pdf

U.S. Department of Defense, *DoD Climate Assessment Tool*, fact sheet, undated.

U.S. Department of Defense, *Department of Defense Strategic Sustainability Performance Plan*, Under Secretary of Defense for Acquisition, Technology, and Logistics, August 26, 2010.

U.S. Department of Defense, *Department of Defense Strategic Sustainability Performance Plan: FY 2016*, Under Secretary of Defense for Acquisition, Technology, and Logistics, September 7, 2016.

U.S. Department of Defense, *Sustainability Analysis Guidance: Integrating Sustainability into Acquisition Using Life Cycle Assessment*, June 24, 2020.

U.S. Department of Defense, Climate 21 Project, homepage, 2021. As of April 10, 2023:
https://climate21.org/

U.S. Department of Defense, *Fact Sheet: 2022 National Defense Strategy*, March 2022a.

U.S. Department of Defense, "DOD, Other Agencies Release Climate Adaptation Progress Reports," DoD News, October 6, 2022b.

U.S. Department of Defense, *2022 National Defense Strategy of the United States of America*, October 27, 2022c.

U.S. Department of Defense, "DoD, GSA Sign MOU to Bring More Environmental Innovators to Federal Marketplace," press release, March 22, 2023.

U.S. Department of Defense Environment, Safety and Occupational Health Network and Information Exchange, homepage, undated-a. As of February 15, 2023:
https://www.denix.osd.mil

U.S. Department of Defense Environment, Safety and Occupational Health Network and Information Exchange, "Annual Reports to Congress," webpage, undated-b. As of February 13, 2023:
https://www.denix.osd.mil/arc/

U.S. Department of Defense Environment, Safety and Occupational Health Network and Information Exchange, "NDCEE Home," webpage, undated-c. As of March 13, 2023:
https://www.denix.osd.mil/ndcee

U.S. Department of Defense Environment, Safety and Occupational Health Network and Information Exchange, "Secretary of Defense Environmental Awards Home," webpage, undated-d. As of February 13, 2023: https://www.denix.osd.mil/awards/

U.S. Department of Defense Environment, Safety and Occupational Health Network and Information Exchange, "Welcome to the Department of Defense Sustainable Products Center," webpage, undated-e. As of February 13, 2023: https://www.denix.osd.mil/spc/

U.S. Department of Defense Environment, Safety and Occupational Health Network and Information Exchange, "DoD Sustainable Products Center: Biobased Grease Demonstration," webpage, undated-f. As of February 13, 2023: https://www.denix.osd.mil/spc/demonstrations/completed/biobased-grease-demonstration/

U.S. Department of Defense Environment, Safety and Occupational Health Network and Information Exchange, "DoD Sustainable Products Center: Biobased Motor Oil Demonstration," webpage, undated-g. As of As of February 13, 2023: https://www.denix.osd.mil/spc/demonstrations/completed/biobased-motor-oil-demonstration/index.html

U.S. Department of Defense Environment, Safety and Occupational Health Network and Information Exchange, *Defense Environmental Restoration Program Annual Report to Congress: Fiscal Year 1995*, 1995.

U.S. Department of Defense Environment, Safety and Occupational Health Network and Information Exchange, *Calendar Year 2004: Executive Order 13148 Annual Report—Department of Defense*, May 16, 2005.

U.S. Department of Defense Environment, Safety and Occupational Health Network and Information Exchange, *2015 Secretary of Defense Environmental Awards: Environmental Excellence in Weapon Systems Acquisition, Small Team—Fairchild Air Force Base*, 2015.

U.S. Department of Defense Environment, Safety and Occupational Health Network and Information Exchange, *2016 Secretary of Defense Environmental Awards: Environmental Excellence in Weapon Systems Acquisition, Large Program—Joint Light Tactical Vehicle Environmental, Safety, and Occupational Health Working Group, Michigan*, 2016a.

U.S. Department of Defense Environment, Safety and Occupational Health Network and Information Exchange, *2016 Secretary of Defense Environmental Awards: Environmental Excellence in Weapon Systems Acquisition, Large Program—KC-46A Program Environment, Safety, and Occupational Health Team*, 2016b.

U.S. Department of Defense Environment, Safety and Occupational Health Network and Information Exchange, *2018 Secretary of Defense Environmental Awards: Environmental Excellence in Weapon Systems Acquisition, Team—Combat Rescue Helicopter Program ESOH Team*, 2018.

U.S. Department of Defense Environment, Safety and Occupational Health Network and Information Exchange, *2021 Secretary of Defense Environmental Awards: Sustainability, Individual/Team—Naval Supply Systems Command, Weapon Systems Support, Pennsylvania*, 2021a.

U.S. Department of Defense Environment, Safety and Occupational Health Network and Information Exchange, *2021 Secretary of Defense Environmental Awards: Sustainability, Individual/Team—Whiteman AFB Environmental Element Sustainability Team*, 2021b.

U.S. Department of Defense Environment, Safety and Occupational Health Network and Information Exchange, *2022 Secretary of Defense Environmental Awards: About the Awards*, fact sheet, 2022a.

U.S. Department of Defense Environment, Safety and Occupational Health Network and Information Exchange, *2022 Secretary of Defense Environmental Awards: Environmental Excellence in Weapon Systems Acquisition, Team—C-130 Program Office and Support Team*, 2022b.

U.S. Department of Defense Inspector General, *Evaluation of the Department of Defense's Efforts to Address the Climate Resilience of U.S. Military Installations in the Arctic and Sub-Arctic*, April 13, 2022.

U.S. Department of Transportation, Federal Highway Administration, "LCA Pave Tool," webpage, updated April 4, 2022. As of February 15, 2023: https://www.fhwa.dot.gov/pavement/lcatool/

U.S. Environmental Protection Agency, "Guiding Principles for Sustainable Federal Buildings," webpage, May 4, 2022. As of February 16, 2023: https://www.epa.gov/greeningepa/guiding-principles-sustainable-federal-buildings

U.S. Environmental Protection Agency, "Sustainable Marketplace," webpage, updated April 19, 2023. As of February 15, 2023: https://www.epa.gov/greenerproducts

"U.S. Navy Starts Alternative Fuel Use," *Maritime Executive*, January 20, 2016.

U.S. Office of Management and Budget, *Department of Defense FY 2021 OMB Scorecard for Federal Sustainability*, 2022.

Vergun, David, "Prototype Aims to Reduce Fuel Use, Improve Tactical Vehicle Performance," DoD News, November 24, 2021.

White House, *National Security Strategy*, October 12, 2022.